FRESHWATER MUSSEL ECOLOGY

FRESHWATER ECOLOGY SERIES

Editor-in-Chief
*F. Richard Hauer, Flathead Lake Biological Station,
University of Montana*

Editorial Board
Emily S. Bernhardt, Department of Biology, Duke University
Stuart E. Bunn, Australian Rivers Institute, Griffith University, Australia
Clifford N. Dahm, Department of Biology, University of New Mexico
Kurt D. Fausch, Department of Fishery and Wildlife Biology,
Colorado State University
Anne E. Hershey, Biology Department, University of North Carolina, Greensboro
Peter R. Leavitt, Department of Biology, University of Regina, Canada
Mary E. Power, Department of Integrative Biology,
University of California, Berkeley
R. Jan Stevenson, Department of Zoology, Michigan State University

University of California Press Editor
Charles R. Crumly

http://www.ucpress.edu/books/pages/11082.html

FRESHWATER MUSSEL ECOLOGY

A Multifactor Approach to Distribution and Abundance

David L. Strayer

UNIVERSITY OF CALIFORNIA PRESS   Berkeley   Los Angeles   London

University of California Press, one of the most
distinguished university presses in the United States,
enriches lives around the world by advancing
scholarship in the humanities, social sciences, and
natural sciences. Its activities are supported by the
UC Press Foundation and by philanthropic
contributions from individuals and institutions.
For more information, visit www.ucpress.edu.

Freshwater Ecology Series, volume 1

University of California Press
Berkeley and Los Angeles, California

University of California Press, Ltd.
London, England

©2008 by
The Regents of the University of California

Library of Congress Cataloging-in-Publication Data

Strayer, David Lowell, 1955–.
    Freshwater mussel ecology : a multifactor approach
to distribution and abundance / David L. Strayer.
        p.   cm. — (Freshwater ecology series; v. 1)
    Includes bibliographical references and index.
    ISBN 978-0-520-25526-5 (cloth)
    1. Freshwater mussels—Ecology. I. Title.
    QL430.6.S8785 2008
    594'.4176—dc22   2007043799

Manufactured in the United States of America
16 15 14 13 12 11 10 09 08
10 9 8 7 6 5 4 3 2 1

The paper used in this publication meets the
minimum requirements of ANSI/NISO
Z39.48-1992 (R 1997) (Permanence of Paper). ♾

Cover illustration: A collection of pearly mussel
shells. Photo by Dr. Chris Barnhart.

CONTENTS

Preface   vii

PART ONE.   THE LABORATORY

1. The Model as Monster   3
2. The Case of Pearly Mussels   9

PART TWO.   THE MONSTER'S PARTS

3. Dispersal   25
4. Habitat   43
5. Hosts   65
6. Food   87
7. Enemies   105
8. Implications for Conservation   113

PART THREE.   MAKING THE MONSTER WALK

9. Three Models for Mussel Ecology   121
10. Is a Comprehensive Model Possible?   143

Literature Cited   157

Index   199

PREFACE

I wrote this little book to try to solve a specific problem and at least shed some light on, if not solve, a very general problem in ecology. The specific problem is understanding how ecological factors determine the distribution and abundance of pearly mussels (Unionoidea). Pearly mussels are widespread and common in fresh waters around the world, and are of special interest because of their conservation problems: human activities have driven dozens of species into extinction and hundreds more into danger of extinction. Furthermore, as dominant suspension-feeders and shell-builders, pearly mussels play important roles in ecosystem functioning. Historically, pearly mussels have supported important fisheries for pearls, mother-of-pearl, and meat, but many of these fisheries have been destroyed by overharvest, habitat destruction, and pollution. Understanding how ecological factors control pearly mussel populations would help us to conserve or restore imperiled populations, understand the functioning of freshwater ecosystems, and manage mussel fisheries. At present, we have many fragments of useful knowledge about pearly mussel ecology, but we do not have a catalog of the pieces of information that will eventually be needed to produce an adequate theory of pearly mussel distribution and abundance, nor do we have a plan for putting all of these pieces together. There is good evidence to suspect that an adequate theory will have to include multiple ecological factors, and I do not think that mussel ecologists have realized just how difficult the problem of integrating multiple factors into a single theory is likely to be.

The general problem that I will address here is that of integrating multiple factors into a single working theory. This problem is pervasive in ecology, although it has not been widely recognized as a problem by ecologists. Although it could be argued that most important ecological variables are under the simultaneous control of multiple factors, most of contemporary ecology still is concerned with the influence of a single factor (or occasionally two factors). As I will show below, and as is apparent from cases such as the controversy over top-down "versus" bottom-up control of plankton populations, integrating multiple controlling factors into a satisfactory working theory is not a trivial problem. I believe that ecologists eventually will have to confront this general problem.

The goal of this book, therefore, is to try to develop a satisfactory, multiple-factor theory to explain the distribution and abundance of pearly mussels (or more fundamentally to assess whether it is even feasible to develop such a theory at all), and to explore the general problem of multi-factor theories. After presenting some basic information on the biology and conservation situation of pearly mussels, I will consider in detail the various factors that might be included in a theory to understand the distribution and abundance of pearly mussels, highlighting information gaps. Finally, I will discuss various ways in which these pieces might be drawn together into a theory, and make suggestions about approaches that might be fruitful. I admit that my attempts at integration are not entirely successful, but I think that they are partly successful, and may inspire others to finish the job.

I wrote most of this book while a visitor in the Laboratoire d'Écologie des Hydrosystèmes Fluviaux at the Université de Lyon 1. I am most grateful to Janine Gibert, Bernhard Statzner, and their colleagues at the University of Lyon for providing a pleasant and quiet place in which to think and work, and for their personal kindness (merci beaucoup!). The Cary Institute of Ecosystem Studies likewise has provided an atmosphere that encourages reflection and synthesis. I thank Bernhard Statzner, Hugh Possingham, Winsor Lowe, and my colleagues at the Cary Institute of Ecosystem Studies for helpful discussions, and Kevin Cummings, Tom Watters, Art Bogan, and Christine Mayer for maintaining the freshwater mollusk bibliography and mussel/host database, which saved me a *lot* of work. Kaustav Roy generously shared his intruiging analysis of the relationship between mussel body size and conservation status (Fig. 6), and Tom Watters and Martin Huehner kindly supplied reports of their mussel surveys for Fig. 11. Lynn Sticker helped check the references while at the same time attending to a leaky roof. I thank the efficient and friendly people at the University of California Press and Michael Bass Associates for their help in turning a pile of loose pages into a book. I particularly appreciate Chuck Crumly's early enthusiasm and advice, and Scott Norton's advice on chapter organization and headings. Finally, I appreciate the support of the Hudson River Foundation, the National Science Foundation, the Nature Conservancy, the New York Natural Heritage Program, and the United States Fish and Wildlife Service for funding my past work on mussel ecology.

PART I

THE LABORATORY

ONE

THE MODEL AS MONSTER

One of the most enduring popular images of science is that of the cinematic mad scientist, scouring a graveyard for body parts from which to build a monster. Dr. Frankenstein is faced with two formidable tasks. First, he and his hunchbacked assistant must gather all of the pieces needed to build a living monster (usually while evading the local constabulary). Even more daunting, he then must find some way to animate the collection of unliving body parts—he must make the monster walk.

Dr. Frankenstein might seem to be a strange role model for scientists, but his method is perhaps a compelling model of reductionist science. In reductionist science, scientists study the various parts of a larger scientific problem or system in the hope that once they understand how all of the pieces work they will be able to put them together and achieve a mechanistic understanding of the whole. Although reductionist approaches are used widely and successfully in many branches of science, including ecology, some critics have claimed that ecological systems are so complex that the reductionist goal of constructing a satisfactory understanding of complex ecological systems from its parts is so difficult so as to be practically unachievable, as well as philosophically unsound (e.g., Peters 1991, Rigler and Peters 1995, Bormann 2005).

In some ways, the task of reductionist science is even more difficult than Dr. Frankenstein's. Most often, a scientific field is not coordinated by a single person—it is as if we have dozens of hunchbacks scouring the countryside for body parts, but no scientist (mad or otherwise) to coordinate the monster-building. Further, unlike Dr. Frankenstein, we don't usually have an explicit plan for animating the parts into a working theory. Although most of ecological science is reductionist in character, scientists rarely explicitly assess whether a reductionist approach to a particular ecological problem is feasible: Can we collect all the parts needed to build a mechanistic understanding of the problem? Can we animate them into a working whole? If these *are* feasible problems, what parts to we need, and how do we best integrate them into a working theory? Here, I explicitly consider what would be required to build a mechanistic understanding of one specific ecological problem: that of predicting the distribution and abundance of freshwater unionoid mussels (Fig. 1); I then assess the feasibility of this enterprise.

This exercise could be done with any group of organisms. I chose unionoid mussels as an example because they are important in conservation and freshwater ecology—thus, there is a specific interest in understanding the factors that limit the distribution and abundance of this particular group of organisms; and because there is no reason to think that unionoids are unrepresentative of the many other groups of organisms that ecologists try to understand or manage; and also because I enjoy working with the group and am familiar with the literature on these animals.

Our current understanding of unionoid ecology is inadequate in three important ways: (1) studies of single factors have not yet led to an adequate understanding of the importance of *any* single factor in nature, even for those factors that have received much study (e.g., habitat, fish hosts); (2) some factors (e.g., food) have not even been seriously considered as controls on natural populations; and (3) the simultaneous influence of multiple controlling factors, although probably important in nature, has not been seriously considered by unionoid ecologists. That is, both Igor[1] and Dr. Frankenstein have work to do. These problems are not unique to unionoid

---

1. Although popularly known as "Igor", the hunchbacked assistant had different names in the various Frankenstein films: "Fritz" in *Frankenstein* (1931), "Ygor" in *Son of Frankenstein* (1936), and "Igor" in *Young Frankenstein* (1974). He didn't appear at all in Mary Shelley's novel. I'll call him Igor.

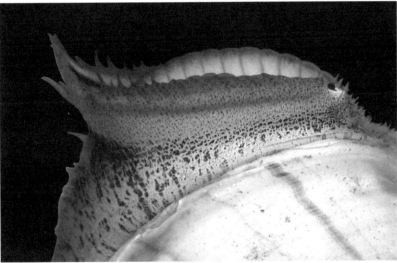

FIGURE 1. Unionoid mussels. Top: Shells of a mixed-species assemblage, showing the lustrous mother-of-pearl for which these animals were prized. Bottom: A gravid female of *Lampsilis cardium* displaying a lure to attract hosts for her larvae. From Chris Barnhart (http://courses.missouristate.edu/mcb095f/gallery).

ecology—I do not believe that there is anything especially pathological about unionoid ecology—but are common to many areas in ecology.

Before proceeding, I want to be clear about what, in my view, would constitute a satisfactory theory of mussel distribution and abundance. Ecology is often said to be concerned with predicting the distribution and abundance of organisms, and I take that goal literally. That is, I think we should strive to make quantitative predictions about the probability of occurrence or abundance of individual species based on environmental and biological variables. An adequate theory should be able to do this well; that is, the theory predictions should be close to actual values of these variables in the field.

Ecologists sometimes seem content to reach alternative goals. For instance, I have the impression that some ecologists would be satisfied if we could just list the factors that determine distribution and abundance of organisms. That is, instead of producing an equation of the form

$$Abundance = 0.15 + (0.5 \times food - 1.66 \times predation)^{-|temperature-20|}$$

they would be happy to say that

$$Abundance = f(food, predation, temperature)$$

Alternatively, much of contemporary ecology is focused on understanding individual processes (e.g., predation, disturbance, competition). This focus has been very fruitful in making generalizations across study systems, and I do not question its value. Nevertheless, both science and society often have compelling needs to know the actual distribution or abundance of individual species. Species, and the ecosystems in which they live, are subject to the integrated, simultaneous influences of multiple processes (cf. Greene 2005). Isolated studies of individual processes or a vague outline of the equation that describes the control of distribution and abundance will not meet the intellectual or practical goals that ecologists and society have set for our discipline. Therefore, I insist that, whatever the value of reaching any other goals, a literal, quantitative prediction of the distribution and abundance of individual species is properly a central goal of ecology.

The objectives of this book are to assess the feasibility of producing such a theory to predict the distribution and abundance of unionoid mussels and, to the extent possible, rough in parts of that theory. I see three possible conclusions of this assessment: (1) we can produce a mechanistic theory (following the reductionist model) that adequately predicts the distribution and abundance of unionoid mussels; (2) we can't produce an adequate mechanistic theory, but we can devise some sort of acceptable alternative that provides useful predictions of unionoid distribution and abundance; or (3) we can't produce adequate predictions by any means and must abandon the problem as scientifically intractable.

I will begin by considering individually the pieces that I think are probably necessary for a working theory of unionoid distribution and abundance, reviewing what we know and what we might ultimately need to know about each part. There are many ways in which to divide up and define these parts, but I will use a five part structure that seems natural to me. The five pieces that I think might be needed to predict unionoid distribution and abundance are dispersal, habitat, fish hosts, food, and predation. Except for the piece on fish hosts, which is needed to account for the unionoids' peculiar parasitic life history, these pieces probably are needed to explain the distribution and abundance of any kind of organism. After I review each of these pieces, I will discuss various ways in which they might be put together. Finally, I will assess our prospects for actually collecting all of the necessary pieces and integrating them into a working theory, a central goal of unionoid ecology.

TWO

THE CASE OF PEARLY MUSSELS

IMPORTANCE OF PEARLY MUSSELS

Pearly mussels of the superfamily Unionoidea (including the families Unionidae, Margaritiferidae, and Hyriidae) are common and widespread in rivers, streams, lakes, and ponds around the world, living on all continents except Antarctica. They can form locally dense populations of >100 animals/m^2 (Fig. 42), and often vastly outweigh other animals in benthic communities, reaching biomasses (not including their shells) sometimes exceeding 100 g dry mass/m^2 (e.g., Hanson et al. 1988, Strayer et al. 1994). Although their roles in freshwater ecosystems have not been fully investigated (Vaughn and Hakenkamp 2001), they can be important suspension-feeders, influencing water chemistry and clarity, and the amount and kind of suspended particles in the water (e.g., Welker and Walz 1998, Vaughn and Hakenkamp 2001). Shell production by unionoids can be of the same order of magnitude as wood production by trees in a temperate forest (Gutierrez et al. 2003, Strayer and Malcom 2007b), providing important physical structure to other organisms (e.g., Chatelain and Chabot 1983, Beckett et al. 1996, Vaughn et al. 2002, Gutierrez et al. 2003). Waste products from mussels can enhance local populations of algae (Vaughn et al. 2007) and macroinvertebrates (Vaughn and Spooner 2006). Thus, the effects of pearly mussels on freshwater ecosystems can be important and pervasive.

Pearly mussels also are economically important to humans. Pearly mussels have been harvested as a source of pearls, mother-of-pearl, and human food since prehistoric times (e.g., Kunz 1898, Morrison 1942, Claassen 1994, Ziuganov et al. 1994, Anthony and Downing 2001, Walker et al. 2001). Freshwater pearl fisheries were one of the reasons that Julius Caesar invaded Britain (Ziuganov et al. 1994), and were one of the most important sources of new capital in 19th century American rural economies (Claassen 1994). Most of these fisheries have disappeared because of overharvest, habitat destruction, or pollution, or because the products they provided have been replaced by other materials (e.g., we now make "mother-of-pearl" buttons and ornaments out of plastic), but regionally important fisheries for shell and pearls still exist (Bowen et al. 1994, Claassen 1994, Neves 1999, Beasley 2001).

EVOLUTION AND CLASSIFICATION OF PEARLY MUSSELS

The major groups of unionoid mussels, their geographic distributions, and the approximate number of species that each contains are now well known (Table 1), but evolutionary relationships among both higher-level taxa and species still are incompletely understood. Traditional classifications based on characters of the shell and soft anatomy have largely been invalidated by molecular studies (e.g., Lydeard et al. 1996, Hoeh et al. 2001, Huff et al. 2004, Campbell et al. 2005). However, molecular data have not yet been collected on enough species to provide a clear picture of evolutionary relationships in the Unionoida.

The order Unionoida usually has been divided into two superfamilies: the Etherioidea, whose larva is a lasidium, and which live in tropical fresh waters around the world; and the Unionoidea, whose larva is a glochidium. The Etherioidea are relatively poorly known ecologically, and will not be dealt with further here. The superfamily Unionoidea contains three families: the Unionidae, by far the largest and most widespread family in the order, the Margaritiferidae, and the Hyriidae.

Members of the Unionidae occur on all of the continents except for Antarctica. Several recent studies (e.g., Lydeard et al. 1996, Hoeh et al. 2001, Campbell et al. 2005) have succeeded in defining several more or less well defined groups of genera ("tribes") within the Unionidae (Table 2), although the placement of all unionid genera in these tribes is not yet

TABLE 1 *Higher-level Classification, Geographic Distribution, and Approximate Species Richness of the Order Unionoida, modified from Cummings and Bogan (2006)*

Taxon	Species Richness	Distribution
Superfamily Unionoidea		
Family Unionidae	707	North and Central America, Europe, Asia, Africa, possibly New Guinea
Family Margaritiferidae	12	North America, Europe, Asia
Family Hyriidae	93	Australia and nearby islands, South America
Superfamily Etherioidea		
Family Mycetopodidae	50	South and Central America
Family Iridinidae	32	Africa
Family Etheriidae	4	South America, Africa, India

NOTE: Members of the superfamily Unionoidea are the subject of the present book.

known. The tribe Lampsilini is usually thought to be the most derived evolutionarily (Campbell et al. 2005).

The Margaritiferidae usually are regarded as primitive relatives of the Unionidae. The family is small and restricted to the Northern Hemisphere, but margaritiferids often are extremely abundant where they occur (cf. Fig. 26). Many margaritiferids use salmonids as hosts. Because of their abundance and the current peril of many of the species, the margaritiferids are perhaps the best-studied of the unionoids (e.g., Ziuganov et al. 1994, Bauer and Wächtler 2001, Huff et al. 2004, and references cited therein).

The evolutionary position of the hyriids continues to be unclear. These animals are common and widely distributed in South America, Australia, New Zealand, and the Pacific Islands. They were originally placed with the etherioids because of their Gondwanaland distribution and anatomical characteristics, but then moved to the Unionoidea when the peculiar lasidium larva of other etherioids was discovered (hyriids have glochidia). More recent analyses have again united the hyriids with the etherioids (Graf 2000, Graf and Cummings 2006), or suggested that they occupy a basal position in the Unionoida (Hoeh et al. 2001, Walker et al. 2006). Like

TABLE 2 *Current Classification of the Superfamily Unionoidea*

Family Unionidae		
Subfamily Ambleminae	*Cyrtonaias*?	Tribe Unionini
Tribe Gonideini	*Dromus*	*Unio*
Gonidea	*Ellipsaria*	Family Margaritiferidae
Tribe Quadrulini	*Epioblasma*	
Cyclonaias	*Glebula*	*Cumberlandia*
Megalonaias	*Lampsilis*	*Margaritifera*
Quadrula	*Lemiox*	Family Hyriidae
"*Quincuncina*" *infucata*	*Leptodea*	
and *kleiniana*	*Ligumia*	Subfamily Hyriinae
Tritogonia	*Medionidus*	*Callonaia*
Uniomerus	*Obliquaria*	*Castalia*
Tribe Pleurobemini	*Obovaria* (part)	*Castaliella*
Elliptio	*Plectomerus*?	*Castalina*
Fusconaia (part)	*Potamilus*	*Diplodon*
Hemistena	*Ptychobranchus*	*Diplodontites*
Lexingtonia	*Toxolasma*	*Paxyodon*
Plethobasus	*Truncilla*	*Prisodon*
Pleurobema	*Venustaconcha*	Subfamily Velesunioninae
Quincuncina (part)	*Villosa*	*Alathyria*
Tribe Amblemini	Subfamily Unioninae	*Microdontia*
Amblema	Tribe Anodontini	*Velesunio*
"*Fusconaia*" *ebena*?	*Alasmidonta*	*Westralunio*
"*Obovaria*" *rotulata*?	*Anodonta*	Subfamily Lortiellinae
Popenaias	*Anodontoides*	*Lortiella*
Tribe Lampsilini	*Lasmigona*	Subfamily Hyridellinae
Actinonaias	*Pyganodon*	*Hyridella*
Cyprogenia	*Strophitus*	Subfamily Cucumerunioninae
	Utterbackia	*Cucumerunio*
		Virgus

NOTE: Modified from Lydeard et al. (1996), Graf (2000), Hoeh et al. (2001), Huff et al. (2004), and Campbell et al. (2005). This classification differs from those widely used in the 19th and 20th centuries, and probably will change as more molecular data become available. The correct placement of the many genera that have not been investigated using molecular methods is not yet clear, so such genera are not included in the table.

the Unionidae, the hyriids have been divided into several subfamilies (Table 2). Although this subfamilial classification hasn't been fully tested with molecular methods, it has received partial support (Graf and Ó Foighil 2000).

The evolutionary relationships among the major groups of the Unionoida still are unclear (see Graf and Cummings 2006 for a critical discussion). In particular, the position of the Hyriidae is unclear (indeed, it is possible that they don't even belong in the superfamily Unionoidea despite possession of a glochidium larva), and the relationships and memberships of the subfamilial groups of the Unionidae and Hyriidae remain to be fully defined and tested. Further molecular data and analyses should help us understand the evolutionary relationships, geographic origins and spread, and the development of biological traits of the unionoid mussels.

The genus- and species-level taxonomy of the unionoids also is in a state of flux. Many familiar genera now appear to be polyphyletic (Baker et al. 2004, Huff et al. 2004, Campbell et al. 2005) and will need to be redefined. At the species or subspecies level, cryptic speciation and geographic differentiation both appear to be common (e.g., Davis 1983, 1984, King et al., 1999, Baker et al. 2004, Jones et al. 2006, Serb 2006), so that traditional views about the limits and internal phylogeographic structure of some species will have to be rethought. This fine-scale differentiation has profound implications for the conservation and management of rare pearly mussels.

The advent of molecular methods and statistical analyses has led to very rapid progress in the areas of unionoid evolution and classification over the past 10–20 years. Although the field is very much in flux, I expect that this rapid progress will continue, and that many of the important questions about unionoid evolution will be satisfactorily resolved in the next few years.

BIOLOGY OF PEARLY MUSSELS

Sexes are separate in most unionoid species, although a few species are normally hermaphroditic (van der Schalie 1970, Kat 1983). Hermaphroditism occurs occasionally in many other species, and apparently can be induced by low population density (Bauer 1987a, Walker et al. 2001). Sperm is shed into the water, taken up by females, and fertilizes the eggs held in the females' gills. The fertilized eggs develop into specialized larvae (glochidia) that are held in the females' gills for weeks to months. The developed larvae are obligate, more or less species-specific parasites of fish (known host relationships were complied by Cummings and Watters 2005).

Although it was once believed that glochidia were simply broadcast into the water to await a chance contact with the proper host, it has become increasingly clear that unionoids use a wide range of sophisticated (sometimes almost unbelievable) methods to get their larvae onto hosts: the females have elaborate moving lures (Fig. 1; Kraemer 1970, Haag and Warren 1997, Haag et al. 1999, Corey et al. 2006), or the glochidia are packaged to resemble fish food (Haag et al. 1995, Hartfield and Hartfield 1996, Watters 1999, 2002, Haag and Warren 2003). Some mussels even catch and hold their fish hosts while infesting them with glochidia (Barnhart 2006)! It is worth noting that at least one species bypasses the parasitic stage altogether (Barfield and Watters 1998, Lellis and King 1998, Corey 2003), simply releasing small juveniles onto the sediments, and this short-circuit may occur in other species (Lefevre and Curtis 1911, Howard 1914). Once the larvae attach to the host, they encyst and transform into juveniles. This parasitic period lasts for several days to several months (Coker et al. 1921, Young and Williams 1984b, Watters and O'Dee 1999), and is the main opportunity for the mussel to disperse. The juvenile mussel falls to the sediment after transformation is complete. Not much is known about the juvenile phase, but most juveniles apparently live an interstitial life, buried in the sediments (Yeager et al. 1994, Sparks and Strayer 1998, Smith et al. 2000). They may deposit-feed as an alternative or supplement to suspension-feeding (Yeager et al. 1994). The juvenile phase lasts for one to a few years (Coker et al. 1921, Jirka and Neves 1992, Haag and Staton 2003), after which sexual maturity is attained and the adults are more or less epifaunal (but see Smith et al. 2000, 2001, Schwalb and Pusch 2007), living at or near the sediment surface. Adults probably are mainly suspension-feeders (this will be discussed in more detail below), and may live for one to several decades (e.g., Bauer 1992, Haag and Staton 2003, Howard and Cuffey 2006).

The very long life span of unionoids may be an adaptation to deal with a highly variable environment in the same way that plants use seed dormancy to deal with temporal variability (cf. Levin 1992). Some authors (e.g., Anthony et al. 2001) have suggested that most estimates of unionoid life spans are flawed, and underestimate actual life spans by a factor of three to five, suggesting that life spans of more than 200 years (!) might be common. All of the specific details of this life history (e.g., seasonal timing, fecundity, size of larvae, age of sexual maturity, maximum life span) vary substantially among (and within) species (Bauer 1994, Watters and O'Dee 1999, Bauer and Wächtler 2001, Haag and Staton 2003). Additional information on the basic biology and ecology of unionoids is available in the reviews of Bauer and Wächtler (2001) and McMahon and Bogan (2001); the review of Coker et al. (1921) of early research is also useful.

CONSERVATION ISSUES

By now, the conservation plight of unionoids has been well described (e.g., Bauer 1988, Bogan 1993, Neves 1993, Williams et al. 1993, Ziuganov et al. 1994, Araujo and Ramos 2000, Master et al. 2000, Young et al. 2001), so there is no need to present a detailed analysis here. Instead, I will briefly review the conservation status of the unionoids and emphasize a few points that may not have received adequate attention. Because unionoids are speciose, highly endemic, sensitive to human impacts, and live in freshwater habitats that often have been treated badly by people, many species have declined severely or gone extinct in modern times. In North America, for example, 37 of the approximately 300 species of native unionoids that were extant in the 19th century are already extinct, and another 105 species are imperiled or critically imperiled (Fig. 2; Master et al. 2000). Thus, unionoids often are said to be the most imperiled of any major group of animals in North America (e.g., Master et al. 2000), and are now the focus of concerted conservation efforts (see USFWS 2007 for an introduction to such efforts in the United States). Similar situations occur on other continents (e.g., Araujo and Ramos 2000, Beasley 2001, Walker et al. 2001, Young et al. 2001, Brainwood et al. 2006, IUCN 2006), although unionoid faunas may be less diverse (e.g., in Europe), or declines may be less well documented or conservation activities less advanced than in North America. It is worth emphasizing that freshwater organisms are generally much more imperiled than terrestrial organisms (Master et al. 2000, Dudgeon et al. 2005, Strayer 2006), and unionoids are more imperiled than most other freshwater organisms, so the conservation status of unionoids is much more precarious than that of the more familiar terrestrial vertebrates and vascular plants (as pressing as their conservation needs may be) (Fig. 2).

Although the focus of conservation activities often is on species extinction, it is important to remember that human impacts thin and destroy populations as well as entire species. It may be helpful to think of

FIGURE 2. Global conservation status of species of unionoid mussels and terrestrial vertebrates from the United States, from Master et al. (2000).

the geographic range of a species as an irregular-shaped piece of cloth, with thickness as well as geographic extent. Human activities wear holes through the fabric (local extinctions) as well as thinning the fabric (diminishment of local populations). Just as in a real piece of cloth, this range-thinning affects the strength and integrity of a species. Figures 3 and 4 show examples of this human-induced thinning at different geographic scales, which has affected unionoid populations nearly everywhere around the world. This thinning of the range leads to (1) a diminished role of the species (or of unionoids in general) in local communities and ecosystems; (2) loss of genetic diversity within species; (3) increased distance and presumably greatly reduced dispersal among remaining populations, which may lead to further losses of populations through metapopulation dynamics (see below); and ultimately (4) increased risk of extinction for the species. Such range-thinning has been documented much less well than range-wide extinctions (although the data exist for such documentation, in many regions), and deserves closer attention from unionoid ecologists and conservationists.

The loss of unionoid species and populations has been highly nonrandom, with losses concentrated in certain taxonomic groups and perhaps in certain functional groups. Thus, some genera have nearly disappeared, while others have shown little response to the human transformation of their habitats (Fig. 5). There have been some attempts to relate this differential loss to the functional traits of the species involved. Some authors have noted that the Anodontini have not fared as badly as other groups of unionoids (e.g., Bates 1962, Metcalfe-Smith et al. 1998). The anodon-

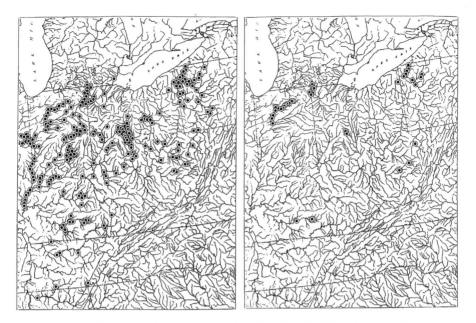

FIGURE 3. Range thinning at a large spatial scale. Historic (left) and current (i.e., ca. 1990; right) range of *Pleurobema clava*, showing substantial range thinning at the regional scale. From Watters (1994a).

tines have several distinctive traits (many anodontines mature quickly, use many species of fish as hosts, have thin shells, and live in many kinds of habitats, including impounded waters), and some authors have singled out specific traits from this list as being responsible for the success of the anodontines. Kaustav Roy (unpublished) has noted that small-bodied unionoids tend to have a much poorer conservation status than large-bodied species (Fig. 6), although there are of course exceptions to this pattern (e.g., the very large and critically imperiled *Margaritifera auricularia* (Araujo and Ramos 2000)). The mechanisms behind this strong pattern are not understood. Figure 4c suggests that species that were widely distributed before human impacts may have suffered proportionately smaller losses than narrowly distributed species. Below, I will show that unionoid species that use many species of fish as hosts are not as likely to be imperiled as those using just a few fish species (Fig. 30). Further analyses such as these on the differential sensitivity of unionoid species to human impacts may lead to helpful insights that can be applied to conservation.

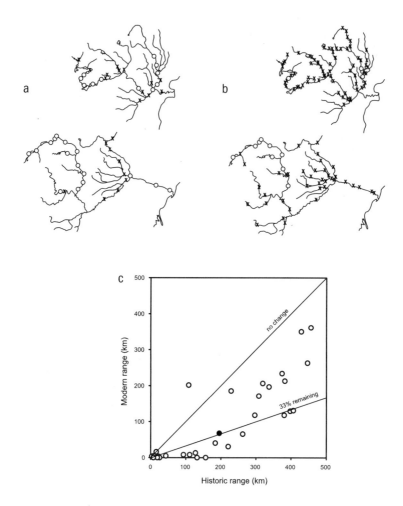

FIGURE 4. Range thinning at the local level. Maps show past (a) and present (b) distributions of *Lampsilis fasciola* in the Raisin and Clinton River basins of southeastern Michigan. Small x's show places where collections were made without finding *Lampsilis fasciola*. Panel (c) shows the degree of range thinning of all 33 species of unionoids from these basins, in terms of km of stream occupied. The black dot in panel (c) is *Lampsilis fasciola*, which is close to the median species in terms of the historic range size and degree of range thinning. Data from Strayer (1979, 1980), McRae et al. (2004), and unpublished records of the University of Michigan Museum of Zoology.

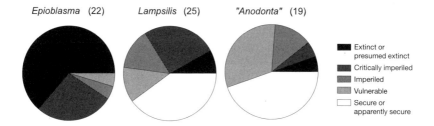

FIGURE 5. Differences in current conservation status of species in selected genera of North American unionoids, based on data of Nature Serve (2005). "*Anodonta* s.l." is a group of genera whose species are now placed in *Anodonta*, *Pyganodon*, and *Utterbackia*. The number of species in each genus is given in parentheses.

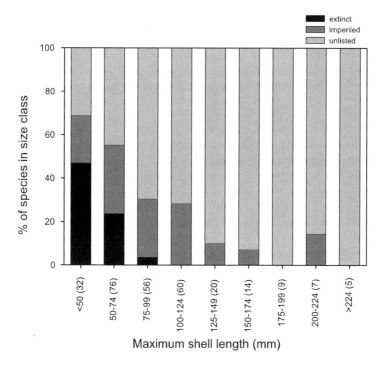

FIGURE 6. The relationship between body size (maximum shell length) and conservation status of North American unionids. Numbers of species in each size class are given in parentheses along the x-axis. Data compiled by Kaustav Roy (unpublished) from various sources, including Neves' (1999) list of extinct species and the USFWS (2007) list of endangered and threatened species.

It is important to remember that the processes that led to losses of unionoid populations and species in the past are not necessarily those that are causing mussel populations to decline today or those that present the greatest threats in the future. For example, construction of large dams on rivers with diverse unionoid communities, and widespread and dreadful water pollution killed many unionoids in the 19th and 20th centuries (e.g., Ortmann 1909, Strayer 1980, Neves et al. 1997, Watters 2000). At present, construction of new large dams has slowed (although of course the existing dams still affect mussels), and point-source pollution has been successfully controlled throughout much of North America and Europe. Thus, there is reason to expect that new construction of large dams and water pollution will be relatively less important in imperiling unionoid populations in the future than they were in the past. Conversely, the zebra mussel (*Dreissena polymorpha*) was introduced from Europe into North America in about 1985, and is still spreading through North American waters (Drake and Bossenbroek 2004). It has already had large effects on North America's unionoids (Strayer 1999b), and may have very large effects in the near future (Ricciardi et al. 1998). Thus, the identity, severity, taxonomic selectivity, and geography of anthropogenic threats to unionoids change constantly, and we must be careful about extrapolating future trends from patterns of past losses.

Finally, the long life span of unionoids and the slow response time of key parts of the ecosystem, e.g., sediment routing through the drainage network (Trimble 1981, Jackson et al. 2005) and nutrient saturation in the watershed (Aber et al. 1998, 2003), mean that the effects of human actions may take many years or decades to be fully expressed as changes in mussel populations. For example, cold water downstream of hypolimnetic-release dams has suppressed mussel reproduction since these dams came into operation in the early to mid-20th century (e.g., Heinricher and Layzer 1999), but long-lived mussel species that recruited before the dams were built still live in these sites. Their populations won't finally disappear until the last of the long-lived mussels dies many decades after the dam that ultimately destroyed the population was built. Similar long lags can follow other human-induced changes to habitat or fish populations (e.g., Kelner and Sietman 2000). Conversely, the full geographic expansion of weedy mussel species of *Pyganodon*, *Anodonta*, *Toxolasma*, and *Uniomerus* into reservoirs built in the 20th century (Bates 1962, Taylor 1984,

Blalock and Sickel 1996, Hughes and Parmalee 1999, Garner and McGregor 2001) probably will take decades or even centuries. Likewise, our present actions, negative or positive, will cast shadows on unionoid populations that reach for decades to centuries. Such time lags substantially complicate analyses of human effects on unionoids, and probably generally lead us to vastly underestimate the effects of human actions on unionoids.

PART 2

THE MONSTER'S PARTS

THREE

DISPERSAL

In the next five sections of the book I will review the five processes—dispersal, habitat, fish hosts, food, and enemies—that I think have the potential to control the distribution and abundance of unionoids. In each of these sections, I will briefly review the state of knowledge about the process, assess the frequency and severity of limitation of mussel populations by the process, try to identify the conditions under which the process is most likely to be limiting, and highlight what I see as critical informational needs.

I do not explicitly address interspecific interactions among coexisting unionoid species. Indeed, except for considering interactions with fish hosts or predators, I do not explicitly consider interactions between unionoids and most other members of the biological communities in which they live (e.g., net-spinning caddisflies, black flies). I am not at all arguing that such interspecific interactions are unimportant in determining the distribution and abundance of mussels, but I will treat their effects implicitly by considering their effects on shared resources.

Broadly speaking, we can think of dispersal as serving two essential functions: (1) dispersal allows a species to move into previously unoccupied areas and thereby expand its geographic range; and (2) dispersal connects the subpopulations within the established range of the species and contributes to the maintenance of unionoid metapopulations. Neither of these types of dispersal by unionoids has received much modern attention.

Dispersal into new regions has at least implicitly been regarded as of central importance in setting the boundaries to the geographic ranges of unionoids. For example, it has been known for more than a century that unionoid distributions form broad zoogeographic regions in which many species co-occur and share more or less congruent range boundaries (Fig. 7; Simpson 1896, Ortmann 1913, van der Schalie and van der Schalie 1950). It has been assumed that the boundaries of these regions usually are set by insurmountable barriers to dispersal. This assumption is probably true in many but not all cases, and is untested.

It seems likely, however, that some range boundaries are set at least in part by climate or other ecological factors. Certainly some range boundaries do not correspond with any obvious barriers to dispersal. The best North American example probably comes from the Canadian Interior Basin. Unionoid species occupy only a small part of this large, climatically harsh region, and their range boundaries do not usually correspond with obvious barriers to dispersal (Fig. 8). There are other examples of range boundaries that are not obviously set by barriers to dispersal. For example, *Fusconaia flava* and *Alasmidonta viridis* are absent from large parts

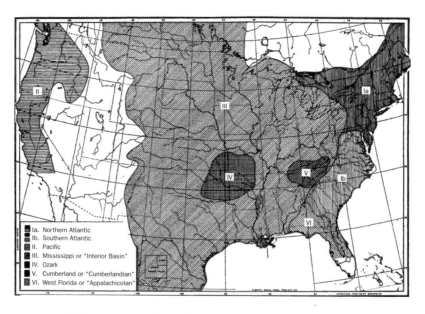

FIGURE 7. Major zoogeographic regions of North America, based on distributions of unionoid bivalves. From van der Schalie and van der Schalie (1950).

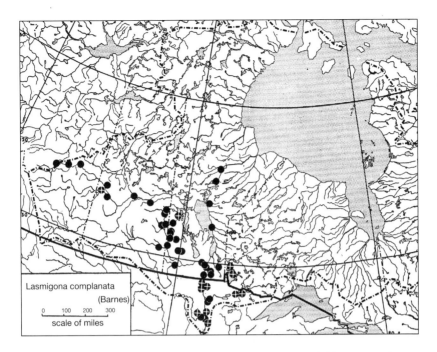

FIGURE 8. An example of a unionoid range boundary that probably is not set by dispersal: the distribution of *Lasmigona complanata* in the Canadian Interior Basin, from Clarke, A.H. 1973. The freshwater molluscs of the Canadian Interior Basin. Malacologia 13: 1-509.

of the upper Ohio River basin in Pennsylvania and New York (Ortmann 1919, Strayer and Jirka 1997), although there does not appear to be any reason that they could not reach this area.

Climate or other ecological factors also may work in concert with inadequate dispersal to jointly set range boundaries. For example, the northern boundary of the Interior Basin zoogeographic province may be set or reinforced by the climatic changes that occur along the northern edge of the Great Lakes. Likewise, the lower elevational boundaries of the Cumberlandian and Ozarkian provinces may be set or strengthened by climatic or ecological changes between these highland areas and the surrounding lowlands.

Despite these observations, there is strong evidence that barriers to dispersal often do set unionoid range boundaries. As argued noted most effectively by van der Schalie (1945; see also Ortmann 1913, van der Schalie

1939, van der Schalie and van der Schalie 1963), unionoid ranges often end at drainage divides, despite the existence of apparently very similar ecological conditions (including populations of the host fish) on the other side of the divide (Fig. 9). It is difficult to interpret such geographic distributions as being anything other than dispersal-limited.

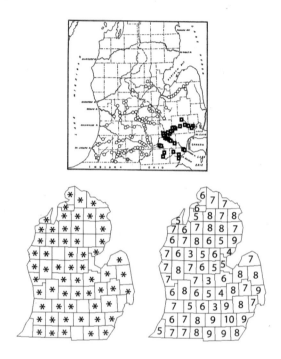

FIGURE 9.   Unionoid range boundaries that probably are dispersal-limited. Top: distribution of *Lampsilis fasciola* (squares) and *Venustaconcha ellipsiformis* (circles) in southern Michigan as of ca. 1940 (modified from van der Schalie 1945). Current distributions of the species are similar, with a few small exceptions presumably as a result of recent introductions. Lower left: distribution of the host fish (smallmouth bass) of *Lampsilis fasciola* in southern Michigan (University of Michigan Museum of Zoology 2005). Asterisks show counties from which smallmouth bass have been recorded; the absences from a few counties probably represent inadequate collecting effort rather than true absences. Lower right: the number of known host species of *Venustaconcha ellipsiformis* that have been reported from each county in southern Michigan (University of Michigan Museum of Zoology 2005). Host fish information from Cummings and Watters (2005), and base map from the Michigan State University Library Digital Sources Center (2005). Ecological conditions are not strikingly different between the eastern and western sides of Michigan.

More direct evidence showing the importance of dispersal limitation comes from cases in which humans have breached drainage divides. The Erie Canal cut through the Alleghenian Divide in 1825, linking the waters of Lake Erie with those of the Mohawk River of the Atlantic Slope. Although conditions in the canal system apparently were not always suitable for unionoids or fish (Daniels 2001), several unionoid species rapidly moved east into the Mohawk basin in the ensuing decades (Fig. 10). Likewise, it appears that the stocking of smallmouth bass into the Potomac

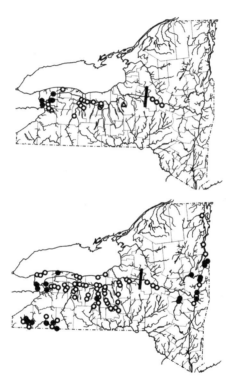

FIGURE 10. Range extensions of the unionoids *Fusconaia flava* (top) and *Pyganodon grandis* (bottom) from the Interior Basin (i.e., Lake Erie) into the Hudson River basin following the opening of the Erie Canal in 1825 (modified from Strayer and Jirka 1997). The heavy bar shows the approximate location of the drainage divide before the canal. Although both species are widespread west of the Alleghenian Divide, they are found east of the divide only along the courses of the Erie and Champlain Canals. (The different symbols show collections from different time periods and are not relevant here). Modified from Memoir 26, New York State Museum, Albany, NY 12230.

basin in the late 19th century carried its unionoid parasite *Lampsilis cardium* into the basin, where it spread widely (Ortmann 1912, Marshall 1917, 1918, 1930) and apparently hybridized with the native *Lampsilis cariosa* (Clayton et al. 2001).

Understanding the importance of dispersal limitation in setting range boundaries is critically important in two management problems: re-establishing unionoid species from parts of their range from which they have been eliminated by human activities, and predicting the responses of unionoid species to climate change. Human activities such as water pollution, damming, and other habitat modification have eliminated unionoid species from large parts of their historical ranges. As these conditions are ameliorated, for example by pollution control or dam removal, it may be possible for unionoids to re-establish populations in these areas, but only if they can disperse into them from other parts of their range. If dispersal is inadequate, then it may be necessary for humans to restock mussels into newly suitable parts of their range to facilitate recovery.

Dispersal rates also will determine the extent to which unionoid species will be able to respond to rapid climate change in the near future. At one extreme, if dispersal is very fast, unionoids will simply follow changing ecological conditions and re-establish their geographic ranges wherever ecological conditions are suitable in the future. At the other extreme, if unionoid dispersal is very slow, then the future range will include only regions that present suitable ecological conditions both now and in all future climates. The actual response of unionoid species will fall somewhere in the large area between these two extremes. Responses to unionoid species to climate change might best be modeled hierarchically, assuming rapid dispersal within drainages and zero dispersal across major drainage divides. Whatever their structure, we will need such models, including accurate estimates of dispersal parameters, to predict the responses of unionoids to climate change. Again, if cross-drainage dispersal severely limits the ability of mussels to adjust their geographic ranges in response to climate change, it may be necessary to consider human intervention.

The second essential function of dispersal is to connect populations within the established geographic range of the species. Unionoids typically live in more or less discrete patches within a drainage basin, whether in lakes or mussel beds in a stream, and populations in different drainage basins within a unionoid species' range are more or less isolated. Dispersal can connect some or all of these separate populations into a metapopulation, providing genetic exchange, rescuing individual populations that decline or disappear, and establishing new populations as suitable habitat is created. Although there has been some interest in analyzing unionoid populations as metapopulations (Vaughn 1993, 1997), little is known about actual dispersal rates within drainage basins or about how such local dispersal contributes to the persistence of local populations. Actual dispersal rates must vary widely across unionoid species, in part because dispersal rates across (and even within) species of host fish vary so widely (Rodriguez 2002). Small benthic fishes (e.g., darters, sculpins) are especially frequent hosts for unionoids (Cummings and Watters 2005). Such fishes may have very limited mobility (Petty and Grossman 2004, McLain and Ross 2005), so dispersal rates of unionoid species that depend on these fishes may be low, at least over long distances. On the other hand, if the mussel uses a mobile or migratory host, its dispersal rate may be very high.

Recent genetic studies of unionoids also suggest that dispersal rates may range widely across species. Some species are highly differentiated genetically within and across drainages (e.g., Nagel 2000, Hughes et al. 2004, Mock et al. 2004, Geist and Kuehn 2005) suggesting low dispersal rates, while other species show much less differentiation (Elderkin et al. 2007), suggesting much higher past dispersal rates.

Thus, we know relatively little about the role of dispersal in limiting the distribution and abundance of unionoid populations, although such information could greatly aid our understanding and management of these animals.

HOW LARGE IS THE EXTINCTION DEBT FROM REDUCED DISPERSAL RATES?

One important application of metapopulation models is estimating the degree to which unionoid species might be affected by human-induced reductions in dispersal rates. This is a special case of "extinction debt" (Tilman et al. 1994), in which the effects of human actions on biodiversity are not fully realized until many years after those actions took place. In this case, the delay may be very substantial because the long life cycle of unionoids will make metapopulation dynamics take decades to centuries to play out.

Many human actions ought to reduce dispersal rates of unionoids. Large dams are absolute barriers to dispersal, and even small dams may block unionoids and fishes (Fig. 11; Watters 1996, McLaughlin et al. 2006). The enormous number of lowhead dams in some parts of the world (e.g., Fig. 14) may have greatly reduced dispersal of unionoids through drainage systems. In addition, pollution, habitat degradation, and reductions in host

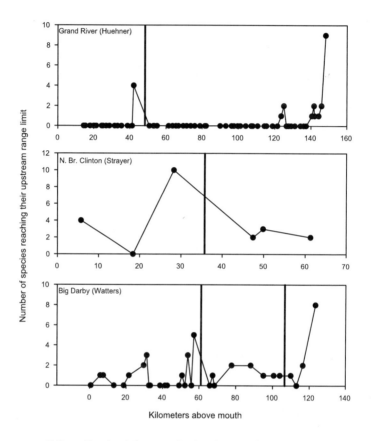

FIGURE 11. Effect of lowhead dams on the distribution of unionid mussels. The graph shows the number of mussel species that reach their upstream range limit at each collecting station. If dams are important, then many species should reach their upper range limits at stations just downstream of dams (thick vertical lines). The analysis includes only sites at which at least 30 living mussels (or 30 living plus "fresh-dead" mussels, in Big Darby Creek) were observed. Based on surveys of Strayer (1980), Watters (1990), and Huehner (1996, 1997, 1999); information on lowhead dams from Ohio Division of Water (2006).

fish populations all may render long reaches of stream unsuitable for unionoids and thereby reduce dispersal rates.

Consider first a very simple two part metapopulation model in which humans affect both the total amount of suitable habitat and the movement rates among patches of suitable habitat. The model can be thought of as applying to a single stream network without dispersal barriers. Treat the network as consisting of a series of N units (e.g., 1-km-long stream reaches). Assume the metapopulation was in equilibrium before human impacts so that

$$\frac{dP}{dt} = b_p - m_p = 0$$

where P is the number of patches that are suitable habitat for the unionoid species, b_p is the "birth rate" of patches (the rate as which unsuitable patches become suitable through natural processes) and m_p is the "death rate" of suitable patches (the rate at which suitable patches become unsuitable as a result of natural processes).

The second unit of this model consists of a simple metapopulation model within the suitable habitats. Consider now just the P patches of suitable habitat. Some fraction of these will be occupied, newly colonized, or abandoned by the mussel species. Following the Levins metapopulation model (Vandermeer and Goldberg 2003), whose predictions are consistent with observed incidence functions for unionoid assemblages (cf. Vaughn 1997), the dynamics of populations in the subuniverse of suitable patches can be described as follows

$$p^* = (m - e)/m$$

where p^* is the equilibrium patch occupancy rate, m is the migration rate (i.e., the proportion of occupied patches that colonize new patches, at arbitrarily low patch occupancy rates), and e is the extinction rate.

Humans probably have had two main effects on this system. First, we have decreased the migration rate m among patches of suitable habitat by imposing barriers between patches of suitable habitat. Second, we have decreased the amount of suitable habitat by increasing m_p and perhaps decreasing b_p. This both reduces the amount of suitable habitat and often increases the distance among the remaining patches of suitable habitat.

Consider first the effects of barriers on reduced migration rates. If the migration rate m is reduced to some fraction f of its original value, then the equilibrium proportion of patches p^* that is occupied is reduced by

$$\frac{e(1-f)}{fm}.$$

Using the fact that $e = m(1 - p^*)$, this can be rewritten as

$$\frac{(1-p^*)(1-f)}{f}.$$

Fig. 12 expresses this loss as a percentage of the number of sites originally occupied for various combinations of p^* and f. Note that the effects of reduced migration rates are disproportionately severe for species that are rare initially (cf. Fig. 4) and that the effects of f and p^* are highly nonlinear (i.e., the contour lines in Fig. 12 are not evenly spaced). This effect alone can be expected to cause a substantial extinction debt on unionoid populations (and perhaps species), and is consistent with the empirical analysis of Fagan et al. (2002) showing that fishes with highly fragmented ranges (and presumably low migration rates) had high rates of local extirpation.

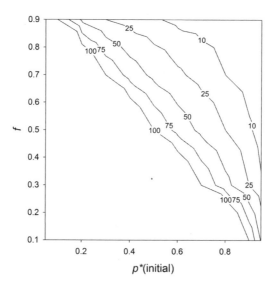

FIGURE 12. Effects of changing within-basin migration rates (m) on a unionoid metapopulation (see text for details). Contours show the percent reduction in equilibrium values of p^*, the proportion of sites with suitable habitat that are occupied by the species; a value of 100 indicates that the species would be eliminated by the reduction in migration rates. The x-axis shows p^* before humans reduced unionoid migration rates, and the y-axis shows the fraction f by which migration rates were reduced.

Where do unionoid populations actually fall on Fig. 12? Based on data such as Strayer (1983: Fig. 6) and Lellis (2001), it appears that unionoids often had high p^* ($\gg$0.5). The degree to which migration rates were reduced is entirely unknown. However, human alterations to streams, especially dams, must have greatly reduced migration rates in many systems. This suggests that many populations fall in the lower right quadrant of Fig. 12 (and also suggests that we critically need actual estimates of reductions in migration rates).

Second, habitat loss per se may increase the size of this extinction debt, either as a direct result of habitat loss or as a consequence of increased distances between remaining populations after some formerly suitable habitats are lost. The direct effect of habitat loss is simply multiplicative with that shown in Fig. 12. That is, if only 40% of the former area of suitable habitat remains and reduced migration rates cause unionoids to occupy only 70% of formerly occupied sites, then their joint effect will cause the unionoid species to occupy just 28% of the sites that it formerly occupied.

The effects of habitat loss on migration rates are more complex to estimate, and depend on the spatial pattern of habitat loss and the shape of the distance-dispersal function. If suitable habitat is lost more or less evenly across the basin, then distances among the remaining patches of suitable habitat will be increased. On the other hand, if the remaining patches of suitable habitat are clustered into one or a few areas (a common pattern in the real world; cf. Figs. 3 and 4), then interpatch distances among remaining patches need not increase at all. Likewise, the effects of habitat loss on the migration rate will be modest if the distance-dispersal function is shallow and severe if it is steep. This function has not been measured for any unionoid, but presumably varies widely depending on the habitats of the fish host, from wide-ranging fishes such as alosids or migrating suckers to immobile fishes such as darters (McLain and Ross 2005) and sculpins (Fig. 13; Petty and Grossman 2004). In any case, habitat loss may substantially increase the extinction debt that arises from the imposition of barriers within a drainage system, especially if the host species disperses poorly.

Remember that this model does not include any effect of humans on the extinction rate of unionoid populations, apart from habitat destruction. Of course, humans may increase extinction rates in ways other than destroying habitat, so humans often put unionoid metapopulations at much greater risk than suggested by Fig. 12.

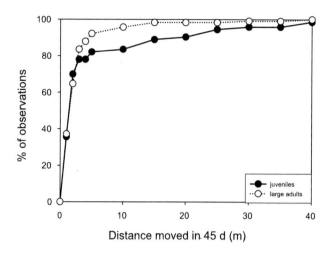

FIGURE 13. Limited dispersal in a stream-dwelling fish. The x-axis shows the distance moved by juveniles and large adults of the mottled sculpin (*Cottus bairdi*) over a period of 45 days in a fourth-order stream in North Carolina. The y-axis shows the percentage of the fish studied that moved a distance less than the distance shown on the x-axis. Note that the period of glochidial attachment often is much shorter than 45 days. From data of Petty and Grossman (2004).

Consider now a larger-scale model in which a unionoid species occupies several drainage systems. It is possible to construct a metapopulation model in which each subpopulation contains of all of the unionoids within a drainage basin, and the metapopulation consists of a series of such basin-wide subpopulations, linked by cross-drainage dispersal. I am thinking here of basins of perhaps 100-1000 km^2 (e.g., the tributaries of Lake Erie, or the tributaries of the Tennessee River), but such a model could be conceived at any scale. Now e is the probability of extinction of the species from an entire subbasin, and m is the migration rate across subbasins. For the sake of illustration, I am guessing that natural unionoid metapopulations like this might have had e~0.001/yr and m~0.02/yr, giving p^*~0.95, although of course we have essentially no data on these parameters.

What might humans do to such a population? First, through habitat destruction, reduced within-basin dispersal, and other impacts, humans will increase within-basin extinction rates. Again, there are no actual quantitative data on the size of this increase, but I think it is reasonable to guess that human-induced increases in within-basin extinction rates might span approximately two orders of magnitude, giving rates of 0.001–0.1/yr. Second, humans probably substantially decrease across-basin dispersal. The larger the scale of the analysis, the larger the effect will be, because of the increased likelihood of including an insurmountable barrier. Again, real data are unavailable, but in view of the enormous number of dams on most river systems (e.g., Fig. 14; Benke 1990, Nilsson et al. 2005, Benke and Cushing 2004) and field data suggesting that even lowhead dams can be effective barriers to mussels (Watters 1996; Fig. 11) and fishes (McLaughlin et al. 2006), as well as the existence of many barriers to dispersal in addition to dams, I think it is reasonable to parameterize a model to examine the effects of reducing across-basin migration by up to approximately two orders of magnitude.

Figure 15 shows the results of this simple cross-basin metapopulation model. Reducing cross-basin dispersal and increasing within-basin extinction each cause severe reductions in patch occupancy by unionoids. The

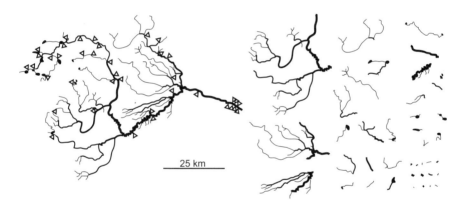

FIGURE 14. An example of how a modern river system (the River Raisin drainage in southern Michigan) has been dismembered by dams. The left panel shows the location of dams (triangles) on the major streams of the basin; an additional 21 dams exist on tributaries too small to be shown on this map. The right panel (drawn to the same scale as the left panel) shows the fragments of the drainage network that have been produced by this dismemberment. Each fragment is separated from other streams in the Raisin basin by at least one dam, and therefore the populations of mussels, fishes, and other animals that it contains have become more or less disconnected from populations elsewhere in the basin. Based on Dodge (1998).

response surface is non-linear (i.e., the contours are unevenly spaced), and is steeper as extinction (the zero isoline) is approached. That is, as metapopulations approach extinction they become increasingly sensitive to reduced cross-basin dispersal and within-basin extinction. If my estimates as to the likely size of human impacts are reasonable, then a large part of plausible parameter space results in the extinction of the metapopulation. In particular, if cross-basin dispersal is reduced by even an order of magnitude, which seems distinctly possible given the degree to which stream systems have been dismembered, then the ultimate size of unionoid metapopulations will be vastly reduced, even if we are able to prevent within-basin extinction rates from falling. This result points out a possible shortcoming of current conservation efforts, which are strongly focused on supporting the viability of local (within-basin) populations.

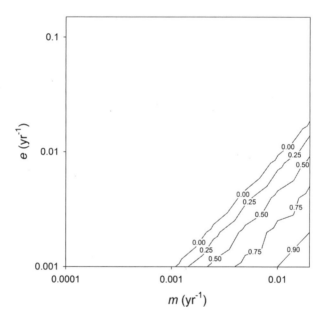

FIGURE 15. Effects of changing across-basin migration rates (m) and within-basin extinction rates (e) in a simple cross-basin metapopulation model (see text for details). Contours show equilibrium values of p*, the proportion of basins with suitable habitat that are occupied by the species. I suggest that pre-industrial unionoid metapopulations in basins of ~100-1000 km² might have fallen somewhere near the lower right of the diagram, with m~0.02/yr, e~0.001/yr, and p^*~0.95.

Thus, whether considered within or across drainage basins, impaired dispersal may cause a large extinction debt in metapopulations of unionoids and other stream-dwelling organisms with limited dispersal abilities. The Levins-type metapopulation models do not offer any insight into the speed of extinctions to be expected from impaired dispersal. I doubt that there are many data on the long-term dynamics of unionoid populations that have been subjected to habitat fragmentation but not to other serious human impacts; such data could be used to parameterize a dynamic model of dispersal impairment in mussel metapopulations. Because of the long life-span of unionoids it may take decades to centuries for this extinction debt to be realized, however.

Both of the simple metapopulation models just discussed are deterministic. Especially in metapopulations where N or p^* are small, stochastic factors can substantially increase the risk of extinction (e.g., Hanski et al. 1996, Lande et al. 1998). Running-water ecosystems are notoriously variable, as are many human impacts (e.g., chemical spills: Guttinger and Stumm 1992, Beck 1996, USFWS 2002), so such stochasticity probably is an important element in real unionoid metapopulations. As a result, the analyses presented here are likely to be conservative, and the extinction debt of unionoids and other poorly dispersing freshwater organisms from human-made barriers is probably larger than these simple models suggest.

Thus, it seems likely that we will need to include dispersal in any satisfactory model of unionoid distribution and abundance, although unionoid ecologists have not written much about formal dispersal models and almost no empirical data exist with which to parameterize such models. Nevertheless, there are at least a few ways in which progress might be made. Further modeling should be useful in defining the circumstances in which dispersal is likely to be limiting (cf. Hughes 2007) and in refining the requirements for empirical measurements. Empirical data on unionoid dispersal might be obtained by observing the recolonization of populations following catastrophic pollution, habitat restoration, dam removal, or experimental defaunation. It might be possible to estimate contemporary cross-drainage dispersal rates from the appearance of new unionoid species outside their historical ranges. Alternatively, genetic studies might be used to estimate both cross-drainage and within-drainage dispersal rates (see Hughes et al. 2004, Mock et al. 2004 and Hughes 2007 for examples),

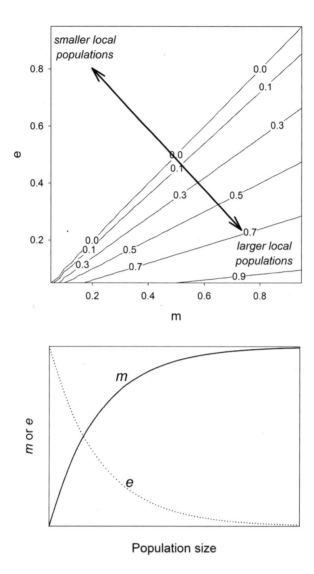

FIGURE 16. Effect of changing the size of local populations in a Levins metapopulation model. Upper: Contours are equilibrium values of p^* (the percentage of sites with suitable habitat that are occupied by a species); m = migration rate, e = extinction rate. Lower: A possible effect of local population size on m and e, entirely speculative.

although it may be difficult to reconcile such estimates with direct measurements of dispersal (Wilson et al. 2004a).

Although dispersal typically has been considered to operate independently of other controlling factors, and sometimes is thought of as hierarchically "above" other controlling factors (i.e., other factors come into play only if dispersal is adequate, e.g., Guisan and Thuiller 2005), it seems likely that there are important interactions between dispersal and other factors that control unionoid populations. For instance, both the extinction rate e and the migration rate m ought to be a function of mussel density within patches (cf. Lande et al. 1998), and thereby depend on controlling factors such as habitat quality, host abundance, predation rates, and food. These interactions have not been studied for unionoid populations; Fig. 16 gives a speculative example of what sorts of interactions might occur.

One final topic that fits broadly under dispersal effects (in this case, the dispersal of gametes) is the possibility of Allee effects occurring at low population densities of mussels. Downing et al. (1993) found that fertilization success of female *Elliptio complanata* living in a Quebec lake was a strong function of local population density. Specifically, fertilization success was highly variable, and often <50% at local population densities below $18/m^2$. Likewise, observations of increased aggregation (Burla et al. 1974, Amyot and Downing 1998) or even male-female pairing (Shelton 1997) of mussels during spawning season suggest that sperm dispersal and concentrations may limit female reproductive success. Nevertheless, other authors have reported no evidence that a large proportion of females are barren, even in sparse populations (Neves 1997, Haag and Staton 2003, and references cited therein), or that fertilization success is a function of local population density (Young and Williams 1984a, Fukuhara and Nagara 1995). Further, the ability of animals living in low-density populations to develop into hermaphrodites (Bauer 1991b, Walker et al. 2001) would tend to mitigate an Allee effect. This subject deserves further attention, because the inability of animals living in sparse populations to reproduce would have important consequences for the viability of such populations. It also would suggest that animals living in sparse parts of a metapopulation would contribute much less to the viability and genetic makeup of that metapopulation than animals living in high-density nuclei.

FOUR

HABITAT

Habitat is probably the first factor to have been thought of as limiting mussel populations. By the time that a scientific literature on unionoid ecology began to develop in the late 19th and early 20th centuries, it was already widely stated that different species of mussels required different habitats (Table 3; e.g., Ortmann 1919, Coker et al. 1921, Baker 1928), and the idea that the amount of suitable habitat limits the size of mussel populations was widely accepted, at least implicitly (e.g., Coker et al. 1921). The notion that the availability of suitable habitat generally limits mussel populations (and corollary ideas, such as the idea that humans destroy mussel populations by degrading habitat quality) is still widely accepted today. Unfortunately, most characterizations of mussel habitat requirements, both historical and contemporary, are vague, untested, and ultimately unsatisfactory. At this point, we do not know how often or under what conditions that habitat limits unionoid populations.

HABITAT REQUIREMENTS OF UNIONIDS

FAILURE OF TRADITIONAL HABITAT DESCRIPTORS

By "habitat," I mean all abiotic factors (including physical disturbance) that affect mussel populations. Traditional habitat descriptions (such as those in Table 3), which were based largely on abiotic characteristics that were

TABLE 3 *Examples of Traditional Descriptions of the Habitat of a Mussel Species* (Cyclonaias tuberculata)

"Found on gravel bars and in mud" (Call 1900)

"From riffles with rather coarse gravel and a rapid flow of water" (Ortmann 1919)

"Found usually on a mud bottom in fairly deep water, 1-2 m" (Baker 1928)

"In the rapids…where there is a coarse gravel and boulder bottom" (van der Schalie 1938)

"Occurs in rivers of various sizes. Found on gravel or mud bottoms" (Clarke 1981)

"Medium to large rivers in gravel or mixed sand and gravel" (Cummings and Mayer 1992)

"In good quality streams and small rivers in sand with a good current" (Watters 1993)

"Lives in large creeks and rivers, often in riffles" (Strayer and Jirka 1997)

"This mussel typically inhabits a gravel/mud bottom, usually in areas of current" (Parmalee and Bogan 1998)

obvious to humans during periods of low water, are unsatisfactory for three reasons. First, critical, quantitative tests of the association between mussel distributions and putatively important factors such as sediment grain size, current speed, water depth, and distance to shore have found that such factors usually are ineffective at predicting the occurrence or abundance of mussels (e.g., Strayer 1981, 1999a, Holland-Bartels 1990, Strayer and Ralley 1993, Strayer et al. 1994, Balfour and Smock 1995, Vaughn and Pyron 1995, Johnson and Brown 2000, Brim Box et al. 2002, Gangloff and Feminella 2007). An example is shown in Fig. 17. Thus, traditional descriptors of mussel habitats do not well describe the places where mussels actually occur.

Second, these habitat descriptions are not portable. Even if mussels do live in a well defined habitat at one site, they often occur in very different habitats at other sites (Fig. 18; Coker et al., 1921; Strayer, 1981). This non-portability suggests that the actual controlling factor is not the putative controlling variable, but rather some unmeasured factor that is variably correlated with this variable. For instance, suppose that mussels don't care about the grain size of the sediment, as long as it is stable during floods. As long as we study only one site, we may find a strong correlation between grain size and mussel distribution, because grain size and stability may be closely related under the specific hydraulic conditions that occur

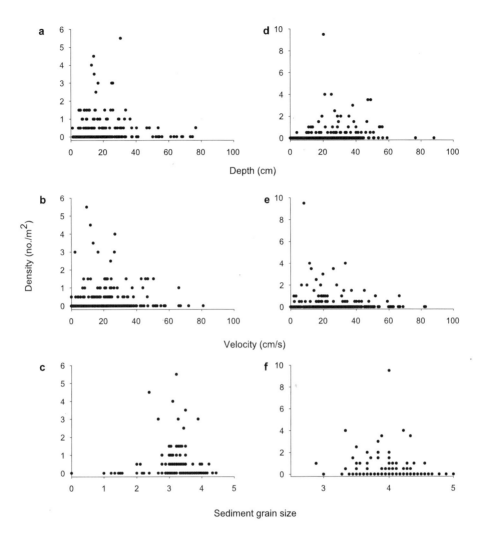

FIGURE 17. Failure of traditionally used habitat descriptors to describe where mussels actually occur. Each point represents the mussel densities and environmental conditions in a single 1 m² quadrat in the Webatuck Creek (panels a–c) or Neversink River (panels d–f) in southeastern New York. None of the linear or quadratic regressions between mussel densities and environmental variables had $r^2 > 0.05$.

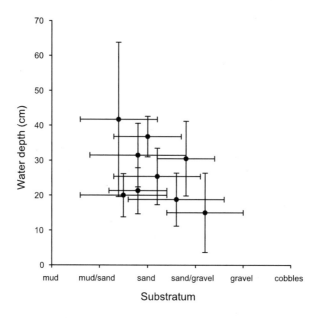

FIGURE 18. Microhabitats (mean±SD) occupied by the unionid *Elliptio dilatata* at nine sites in small streams in southeastern Michigan. From data of Strayer (1981).

at this site. Once other sites with other hydraulic conditions are studied, the relationship between grain size and mussel distribution will break down, because different-sized sediments are stable at the different sites.

Third, even when habitat descriptions are effective at describing mussel distributions across a range of sites, they may give little insight into actual controlling factors. Probably the best example of this is the well known association between mussel distribution and stream size. In many regions, mussel species are restricted to streams within a given size range (Table 4; Baker 1922, 1926, van der Schalie 1938, Strayer 1983, 1993, Haag and Warren 1998). This information is certainly useful when planning surveys or selecting sites for reintroductions. However, mussels presumably don't respond to stream size per se, but to one or more of the many abiotic factors (e.g., temperature, current speed, fertility, organic matter quality, etc.) or biotic factors (e.g., fish distribution, plankton quantity and quality, predation risk, etc.) that vary with stream size. Similar problems plague known associations between mussel distributions and other broad habitat factors such as surface geology (Strayer 1983), hydrology (Di Maio

TABLE 4 *Abundance of Mussel Species in Streams of Different Sizes in the Huron River Basin in Southeastern Michigan*

Species	Brooks	Small creeks	Large creeks	Small rivers	Medium-sized rivers	Fairly large rivers	Large rivers
Lasmigona compressa	C	A	C	C	r		
Anodontoides ferussacianus	r	A	C	C			
Alasmidonta viridis		A	A	r	r		
Strophitus undulatus		C	C	C	C	r	
Elliptio dilatata			C	A	A	A	r
Villosa iris		r	C	A	A	C	r
Alasmidonta marginata			r	C	C	r	r
Pyganodon grandis			r	r	r	r	r
Lampsilis siliquoidea			r	r	r	r	r
Lampsilis cardium				A	C	r	r
Epioblasma triquetra				r	C	r	
Ptychobranchus fasciolaris				A	C	C	r
Lampsilis fasciola				A	C	C	r
Lasmigona costata					C	r	r
Ligumia recta					r	r	r
Utterbackia imbecillis					r	r	C
Cyclonaias tuberculata						A	C
Lasmigona complanata							C
Actinonaias ligamentina							C
Villosa fabalis							C
Fusconaia flava							C
Toxolasma parvum							C

NOTE: r = rare, C = common, A = abundant. Modified from van der Schalie (1938).

and Corkum 1995), and riparian vegetation (Morris and Corkum 1996) for which mechanisms are not known. Although such purely empirical habitat associations are useful, it often would be more useful to understand the mechanisms that underlie these relationships.

One possible alternative to traditional approaches to mussel habitat is to adopt a more functional, "mussel's eye" view of habitat. That is, perhaps we should begin with a list of what a mussel needs from its habitat rather than a list of what we first notice when we visit a stream at low water. Table 5 is a preliminary attempt to develop such a functional definition of habitat. The few studies that have been done of these functional habitat attributes are promising, as shown by the following review of the elements that I've included in Table 5.

LOW SHEAR STRESSES FOR JUVENILE SETTLEMENT

When a juvenile mussel falls from a fish host, it must be able to stay on the sediment long enough to establish itself without getting swept away by excessive current or turbulence. I have not seen any direct studies of the abilities of juveniles to settle under different conditions, although it would seem possible and fruitful to do such work in laboratory flumes. Some authors have interpreted field data as showing that juvenile settlement is impossible when current speed or shear stress is high. Thus, Payne and Miller (2000) noted that *Fusconaia ebena* had just two successful year-classes between 1985 and 1996 in the lower Ohio River. Those two year-classes were established in years when high spring flows (which may have encouraged strong spawning runs of the host fish) were followed by unusually rapid declines in river flow. Payne and Miller suggested that these relatively placid flows are needed to allow juveniles to settle. Likewise,

TABLE 5 *Proposed Functional Characteristics of Suitable Mussel Habitat*

Allows juveniles to settle (shears are not excessive during juvenile settlement)
Provides support (soft enough for burrowing, firm enough for support)
Is stable (stays in place during floods, no sudden scour or fill)
Delivers food (sediment organic matter for juveniles, current provides suspended food to adults)
Delivers essential materials (oxygen, calcium, etc.)
Provides favorable temperatures for growth and reproduction
Provides protection from predators (interstitial juveniles)
Contains no toxic materials

Layzer and Madison (1995), Hardison and Layzer (2001), and Myers-Kinzie et al. (2002) noted negative correlations between shear stress and mussel density or occurrence, which they suggested was the result of negative effects of high shears on juvenile settlement. Nevertheless, other mechanisms could be responsible for a negative correlation between shear stress and mussel density. Indeed, because different species of mussels settle at different seasons—and therefore at different shear stresses—it seems unlikely that limitation of settlement by excessive shear would produce the commonly observed pattern in which all mussel species have similar spatial distributions.

SUPPORTIVE SEDIMENTS FOR BURROWING

Biologists have long suggested that sediments must be firm but penetrable to suit unionoids, but there have been almost no attempts to quantify this need. Using a penetrometer, Johnson and Brown (2000) showed that *Margaritifera hembeli* occurred more often in compact sediments than in soft sediments; they suggested that sediment compaction was a proxy for sediment stability. Lake-dwelling unionoids usually are absent from very soft sediments in deep water (e.g., Headlee 1906, Cvancara 1972, Ghent et al. 1978, Strayer et al. 1981, Hanson et al. 1988). This absence often has been blamed on the sediments being too soft to support mussels (e.g., Headlee 1906, Ghent et al. 1978), although other limitations such as cold hypolimnetic temperatures, low dissolved oxygen, and low food concentrations may be important as well (Cvancara 1972). Hastie et al. (2003) suggested that changing hydrology in small Scottish streams was washing out fine sediments capable of providing burrowing sites for mussels, and thereby reducing recruitment of *Margaritifera margaritifera*. Strayer and Ralley (1993) found that the percentage of sediment that was penetrable was a useful predictor of mussel distribution in a stony New York river, although it accounted for only a small percentage of the variance in mussel distribution. Thus, both excessively soft and excessively hard sediments probably often limit the spatial distribution of mussel populations. It should be possible to quantify the kinds of sediments that are suitable for unionoid burial by using tools such as penetrometers in field surveys and by conducting behavioral studies in the laboratory; I suspect that such studies will show that most species can tolerate a wide range in sediment penetrability.

SEDIMENTS THAT ARE STABLE DURING FLOODS BUT WET DURING DROUGHTS

Sediment stability would seem to be of paramount importance to unionoids living in running waters. These animals are long-lived (usually >10 y) and move too slowly to escape floods (maximum speeds on the order of m/d; Bovbjerg 1957, Young and Williams 1983, Balfour and Smock 1995, Amyot and Downing 1998), but occupy one of the most chronically unstable habitats on earth. Stream sediments typically move during floods every year or two (Leopold et al. 1964, Gordon et al. 1992). It is thus natural to hypothesize that unionoids might occur only on parts of the stream bed that are especially stable. Several studies support this hypothesis (Fig. 19; Vannote and Minshall 1982, Young and Williams 1983, Layzer and Madison 1995, Strayer 1999a, Johnson and Brown 2000, Hastie et al. 2001, Howard and Cuffey 2003, Morales et al. 2006, Gangloff and Feminella 2007). These stable patches may be so well protected that unionoids are unaffected by even major floods (Miller and Payne 1998, Strayer 1999a, Hastie et al. 2001). It is beginning to appear that sediment stability may frequently be a necessary requirement for a stream habitat to be suitable for unionoids. However, recognizing stable habitats (without waiting for a big flood to show where they are), and identifying the degree of stability that is required for various species are likely to be difficult problems. Hydraulic models may be a useful tool (Fig. 20; Lamouroux et al. 1992, Lamouroux and Capra 2002, Howard and Cuffey 2003, Lamouroux and Jowett 2005, Morales et al. 2006).

It goes without saying that areas of stream bottom that are stable at high flows must also usually be under water at low flows (or *very* near to habitats that are under water at low flows) to be suitable for mussels. Thus, areas just downstream of peaking hydropower dams often either are too torrential at high flows or too dry at low flows to support mussels (e.g., Layzer et al. 1993). Even without the effects of peaking hydropower, mussels may be squeezed between midchannel areas too unstable or torrential to be hospitable in high water and nearshore areas that dry out during droughts (Miller and Payne 1998, Gagnon et al. 2004, Golladay et al. 2004).

Abrupt changes in sediment or water regimes induce lateral or vertical instability in stream channels (Leopold et al. 1964, Brookes 1996). Human activities such as forest clearing, row crop agriculture, urbanization, dams (and dam removal), water diversions, riparian habitat destruc-

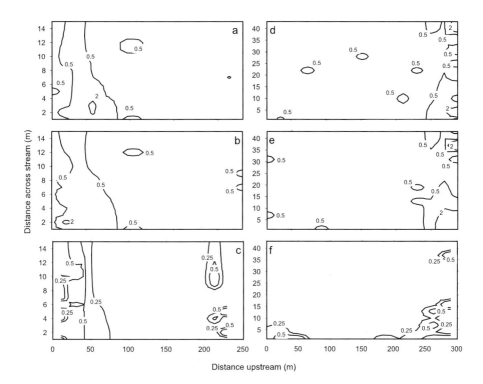

FIGURE 19. Relationship between sediment stability and mussel distribution in two New York streams. The left-hand panels (a-c) show data from Webatuck Creek and the right-hand panels (d-f) show data from the Neversink River. Panels (a) and (d) show mussel densities (number/m^2) the summer before a 5- to 6-year flood, and panels (b) and (e) show mussel densities the summer after the flood. Note the lack of change. Panels (c) and (f) show sediment stability, measured as the probability that a marked stone would stay in place through the flood. From data generated by Strayer (1999a).

tion, and instream gravel mining cause large changes in the amounts of water and sediment moving down stream channels (e.g., Trimble 1981, Hartfield 1993, Waters 1995, Brookes 1996, Brim Box and Mossa 1999, Doyle et al. 2003). These activities are widespread and presumably have large effects on sediment dynamics in running waters throughout much of the world. To the extent that sediment stability is important to mussels, these human activities may have caused widespread harm to mussel populations.

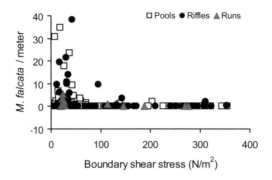

FIGURE 20. Mussel density at various sites along the Eel River, California, as a function of modeled shear stress during the 5-year flood. From Howard and Cuffey (2003).

CURRENTS THAT DELIVER FOOD

I will discuss the food requirements of unionids below, but note briefly here that the characteristics of the habitat may strongly influence delivery of food to mussels. Current speed may affect both the filtration rates of mussels and the development of a food-depleted boundary layer above dense mussel beds, thereby affecting the rate of food acquisition and ultimately mussel growth and fecundity. These effects have been demonstrated to occur in marine bivalves (Wildish and Kristmanson 1997) and zebra mussels (Karatayev et al. 2006). Although Coker et al. (1921) long ago suggested that unionids living in lakes grow more slowly than those living in running waters because food is supplied more slowly in still-water habitats, this idea seems not to have been explored for unionids. Based on marine work (Fig. 21), we can expect growth of adult unionids to be maximized at intermediate current speeds, where the current is fast enough to prevent local food depletion, but not too fast or turbulent to interfere with mussel feeding. It may be relevant that Bolden and Brown (2002) reported slower growth rates of translocated *Margaritifera hembeli* in pools than riffles, perhaps as a result of poor food delivery in pools. Likewise, the quality and quantity of food delivered to interstitial juveniles will be a strong function of interstitial flow rates and directions (i.e., upwelling vs. downwelling; permeable vs. impermeable sediments). The interaction between habitat and food delivery to juvenile unionids appears to be entirely unexplored.

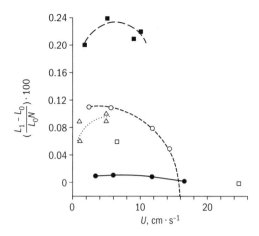

FIGURE 21. Growth of the marine giant scallop (*Placopecten magellanicus*) as a function of water velocity in a series of month-long experiments. Growth is highest at intermediate current speeds. Different curves and symbols represent different times of year. From Wildish and Kristmanson (1997), reprinted with the permission of Cambridge University Press.

ADEQUATE OXYGEN AND CALCIUM

The habitat must of course provide essential materials other than food to unionoids. Perhaps chief among these are oxygen for respiration and calcium for shell growth. Adult mussels may be relatively insensitive to low oxygen. They are able to maintain normal metabolism even at levels of dissolved oxygen as low as 1 mg/L, and can tolerate even complete anoxia for as long as several weeks by simply closing their shells (McMahon and Bogan 2001). Nevertheless, these episodes may reduce mussel growth or cause females to abort glochidia (e.g., Aldridge and McIvor 2003), and longer periods of low oxygen kill mussels. In contrast, juvenile mussels may often be limited by inadequate oxygen. The few studies that have been done on the oxygen needs of juveniles (Dimock and Wright 1993, Sparks and Strayer 1998, Dimock 2000) suggest that their behavior, physiology, and survival may be affected at much higher concentrations than those that affect adults. More importantly, dissolved oxygen concentrations within the stream or lake sediments where juvenile mussels live are much lower (often ~90% lower) than those in the overlying water (e.g., Buddensiek et al. 1993, Strayer et al. 1997). The widespread and large increases in loading of organic matter, nutrients, or fine sediments to lakes and streams may have decreased interstitial oxygen concentrations enough to affect juvenile mussels in many bodies of water. Pollution from sewage and other labile organic matter was widespread and severe throughout much of the developed world in the 19th and 20th centuries (and is still widespread in many less developed parts of the world), and led to hypoxia or anoxia in many streams and rivers (e.g., Hynes 1960). Such episodes of low oxygen killed many fish and presumably many mussels as well. Although oxygen depletion from human activities has largely been controlled in the developed world, it still occurs on occasion. More importantly, mussel communities that were eliminated by past episodes of low oxygen probably have not fully recovered in many places because of inadequate dispersal from remaining source populations through highly fragmented drainage systems.

Although pearly mussels have massive shells of calcium carbonate, they can survive at surprisingly low concentrations of dissolved calcium, through a combination of efficient calcium uptake from food and water and a shell structure that resists dissolution (McMahon and Bogan 2001). Some species of unionoids are able to prosper at ambient calcium concentrations <5 mg/L (e.g., Rooke and Mackie 1984, Huebner et al. 1990). However, calcium concentrations fall below 1 mg/L in the softest fresh waters, and these low concentrations must limit unionoid populations. It also seems possible that juveniles of at least some species might have relatively high calcium requirements and therefore be restricted to calcium-rich waters, although the calcium requirements of juvenile unionoids seem not to have been studied.

MODERATE TEMPERATURES

The temperature of the habitat must be suitable for mussel survival, growth, and reproduction. Both high temperatures and low temperatures may be harmful. High temperatures kill mussels outright, and may be responsible for the deaths of many mussels in droughts (Golladay et al. 2004, Gagnon et al. 2004). High temperatures also decrease the length of time for which glochidia are viable (e.g., Zimmerman and Neves 2002) and increase mortality from other stressors (e.g., Jacobson et al. 1997). Growth is very slow or zero at low temperatures (Hruška 1992, Beaty and Neves 2004), and may decrease at high temperatures as well because of high respiration rates (Huebner 1982, Myers-Kinzie 1998), although analyses of shell growth in *Margaritifera margaritifera* showed that growth was highest at the highest temperatures (Schöne et al. 2004). Further, Hastie et al. (2003) suggested that high temperatures increase recruitment of juvenile *Margaritifera margaritifera* in the British Isles by increasing growth rates of glochidia on fish. Both Chamberlain (1931) and Bauer (1992) noted that growth rates of mussels (in the Northern Hemisphere) were greater in the south than in the north, presumably because of the higher temperatures and longer growing seasons in the south. Bauer (1992) made the important observation that these higher growth rates were correlated with shorter life spans and ultimately a lower rate of population increase, so that higher somatic growth rates do not necessarily translate into higher population growth rates. Generally, high temperatures speed up development (e.g., Dudgeon and Morton 1984, van Snik Gray et al. 2002, Hastie and Young 2003, Steingraeber et al. 2007). Temperature also affects the timing of life history events (Hastie and Young 2003) and thereby presumably determines the match between mussel life-stages and the suitability of habitat conditions (e.g., flow rates for settlement) and availability of hosts. There has been some concern (e.g., Hastie et al. 2003) that global warming may desynchronize mussel life histories from availability of migratory hosts.

Temperature affects reproductive success. Reproduction can be stopped completely by low temperatures, such as those downstream of hypolimnetic-release dams (Heinricher and Layzer 1999). Intriguingly, Roberts and Barnhart (1999) found that glochidial transformation was most successful at low temperatures, perhaps because fish immune function is suppressed at low temperatures. The influence of temperature on unionoids is thus pervasive and complicated, so it is not easy to specify the optimal thermal regime for a unionoid.

REFUGE AGAINST PREDATION

The physical structure of the habitat may provide refuge against predators. Juveniles buried in the sediments may be protected against epibenthic predators such as fish and crayfish. If this interstitial habitat is missing or is rendered unsuitable by human activities, juvenile mussels may be exposed to increased predation rates (cf. Sparks and Strayer 1998). I do not know if this refugial aspect of habitat is ever important to unionoids.

LOW TOXICITY

Finally, the habitat must not contain materials that are toxic to unionids. Most materials of concern are of human origin, although toxic levels of some materials such as ammonia and heavy metals are produced occasionally by natural processes. Unionoids have been eliminated from many places by anthropogenic toxins; major culprits (other than anoxia from domestic and industrial wastes) probably include acid mine drainage (Ortmann 1909, Neves et al. 1997), ammonia (Augspurger et al. 2003), chlorine and chlorination by-products (Goudreau et al. 1993), heavy metals (Naimo 1995), spills of various industrial chemicals (e.g., Crossman and Cairns 1973, Sparks et al. 1999, USFWS 2002), and perhaps synthetic pesticides (Conners and Black 2004) and polycyclic aromatic hydrocarbons (Weinstein and Polk 2001). The influence of anthropogenic toxins on unionoids is the subject of a very large body of literature; for an introduction, see Fuller (1974), Goudreau et al. (1993), Naimo (1995), Weinstein and Polk (2001), Augspurger et al. (2003), Newton (2003), and Conners and Black (2004). While progress in pollution control in developed countries has vastly reduced inputs of toxic substances to fresh waters, toxins probably still have important effects on many unionoid populations. Three especially worrisome classes of pollutants are unionized ammonia, toxic materials with a high affinity for sediments, and endocrine disruptors.

Ammonia is produced from the decomposition of organic matter (and perhaps by dissimilatory reduction of nitrate to ammonia (DNRA); e.g., Kelso et al. 1997, Burgin and Hamilton 2007), and is usually the predominant form of inorganic nitrogen when oxygen is scarce or absent (Wetzel 2001). Ammonia is therefore usually much more abundant in sediments than in the overlying water (Fig. 22). It exists in two forms, the ammonium ion NH_4^+, and unionized ammonia NH_3; the balance between these forms depends on pH and temperature (Fig. 23; Emerson et al. 1975). Unionized ammonia is very toxic to mussels; the 96-hr LC_{50} for juvenile mussels is just at 40–280 μg/L (Augspurger et al. 2003, Newton 2003, Mummert et al. 2003).

There are reasons to think that interstitial ammonia concentrations may have risen worldwide in the last century, and may now often reach these toxic levels. Sediment ammonia should have risen in response to several factors. Decomposition rates should have risen in response to increases in inputs of organic matter from organic pollution or increased autochthonous

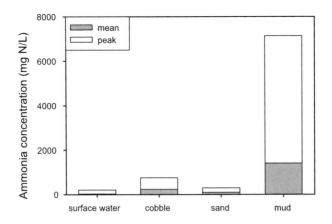

FIGURE 22. Mean and peak ammonia concentrations measured in the surface water and in interstitial habitats in three kinds of sediment in the Pembina River, Alberta. Measurements were taken weekly during the ice-free season for three years. From Chambers et al. (1992).

production of organic matter from enhanced nutrient loading. Large global increases in nitrogen loading to fresh waters (e.g., Vitousek 1994, Vitousek et al. 1997) should have increased the nitrogen content of organic matter and favored DNRA. Finally, ammonia should have increased because of large, widespread increases in inputs of silt and clay (Waters 1995, Brim Box and Mossa 1999), which reduce sediment permeability and thereby reduce interstitial oxygen concentrations. Ammonia toxicity should be most severe when pH and temperature are high, conditions that lead to a high proportion of unionized ammonia (Fig. 23). All of these conditions are especially likely to occur during summer low-flow periods in unshaded agricultural streams of the Midwest, most of which are highly alkaline and have received large inputs of nitrogen and fine sediments (Fig. 24). Thus, excessive ammonia may be an important cause of recent catastrophic declines in unionoids in this region (e.g., Howells et al. 1996, Poole and Downing 2004). High ammonia or low dissolved oxygen may also be responsible for recruitment failures of *Margaritifera margaritifera* in streams with highly compact sediments that are clogged with fine particles (Geist and Auerswald 2007).

Many pollutants (e.g., many metals, organochlorine pesticides, polychlorinated biphenyls, polyaromatic hydrocarbons) are only sparingly

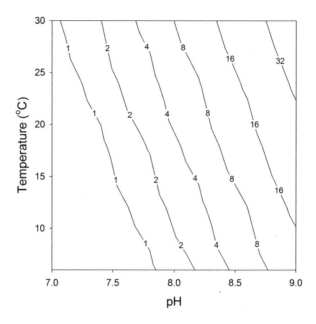

FIGURE 23. Percentage of total ammonia that exists in the unionized form, as a function of pH and temperature. From Emerson et al. (1975).

soluble in water, and therefore concentrate onto sediments. Concentrations of such substances may be orders of magnitude greater in sediments than in the overlying water. Unlike more soluble contaminants, these substances are not easily washed out of ecosystems, and may persist long after the source of the contamination has been eliminated. Because juvenile unionoids live inside sediments and deposit-feed, they may be exposed to much higher concentrations of these toxins than adult mussels or the fish and plankton that are typical subjects of ecotoxicological studies (Naimo 1995). Relatively little work has been done on the effects of sediment-associated toxins on juvenile unionoids, but it seems likely that such substances are preventing the recovery of unionoid populations at sites where sediments still carry the legacy of past pollution.

Finally, endocrine disruptors are chemicals that mimic natural hormones and can disrupt normal reproduction and physiology of animals. Many common environmental contaminants can act as endocrine disruptors, including human or agricultural pharmaceuticals, organochlorine pesticides,

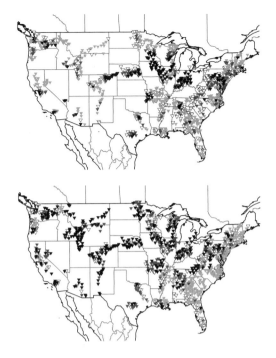

FIGURE 24. Water chemistry in streams, rivers, and lakes of the United States. The upper panel shows mean concentrations of nitrate plus nitrite (typically the dominant forms of inorganic nitrogen): white = <0.21 mgN/L, light gray = 0.21-0.67 mgN/L, dark gray = 0.67-2.18 mgN/L, black = >2.18 mgN/L. The lower panel shows mean values of pH: white = <7.3, light gray = 7.3-7.68, dark gray = 7.68-7.95, black = >7.95. High levels of unionized ammonia are most likely to occur in waters with both high inorganic nitrogen and high pH (dark triangles in both panels). From data of NAWQA (2006); where multiple sampling sites occurred within a small area, I plotted the most frequent value for that area.

tributyltin antifouling paints, and breakdown products of detergents. These chemicals are now widespread in surface waters (e.g., Kolpin et al. 2002) and may have effects in minute quantities. Such substances may interfere with normal reproduction of unionoids and their fish hosts. For instance, endocrine disruptors in sewage effluent caused individuals of *Elliptio complanata* to turn into females (Blaise et al. 2003) and induced spawning (Gagne et al. 2004). Experimental addition of endocrine disrupters at environmentally relevant concentrations disrupted fish reproduction so severely that a fish population disappeared from a Canadian lake (Kidd et al. 2007). It is not yet clear if endocrine disruptors affect the distribution or abundance of unionoid populations through either direct effects or effects on host populations.

We know very well that past episodes of pollution have had major effects on mussel distribution and abundance, but despite a very large body of work on the effects of toxins on unionoids, our knowledge about the extent to which toxins limit unionoid populations is still incomplete. Residual contamination from past pollution continues to limit unionoid populations (e.g., Henley and Neves 1999), as do unexpected spills of toxins (e.g., Crossman and Cairns 1973, Sparks et al. 1999, USFWS 2002). Nevertheless, we know little about the toxicological environment experienced by juvenile unionoids (except that it usually is far more toxic than the overlying water), or about the sensitivity of juveniles to toxins. A lot of the toxicological literature is concerned with defining lethal concentrations of toxins, rather than with sublethal effects on growth, reproduction, and behavior, or interactions between toxins, or between toxins and other controlling factors. Finally, toxicological results haven't often been put into a demographic context, except when the toxin is so deadly that it wipes out the population entirely.

CHALLENGES FOR A FUNCTIONAL APPROACH TO HABITAT
I think that a functional approach to unionoid habitats has promise, although it is not yet clear that this approach will ultimately provide a fully adequate description of unionoid habitats, for several reasons. The variables listed in Table 5 may be the wrong list. A particular problem is that it may be difficult to translate the abstract requirements listed in Table 5 into variables that can readily be recognized or measured in nature. Nevertheless, I would suggest that it may be possible to identify suitable unionoid habitat as areas in which sediments are penetrable and stable for (on average) at least the generation time of the species; concentrations of interstitial unionized ammonia (or human-made pollutants) are never toxic; the temperature regime is warm enough to allow growth and reproduction, but not hot enough to kill or stress mussels; and the current isn't too fast or so turbulent that it interferes with juvenile settlement or adult feeding (Fig. 25). This is a rather different list than the habitat descriptions given in most regional guides or conservation recovery plans. It remains

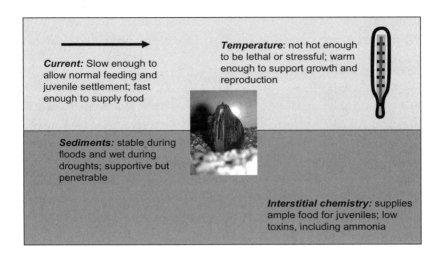

FIGURE 25. Proposed key attributes of suitable unionoid habitat.

to be seen whether a list such as this can be used to produce a numerical measure of habitat suitability, or whether we will have to be content with a binary (suitable vs. unsuitable) categorization of mussel habitat.

CLIMATE

One final habitat issue worth mentioning is the curious lack of attention paid to climate as a limit on unionoid distribution and abundance. Climate is one of the most frequently invoked limits to the abundance and especially the distribution of organisms (e.g., Brown and Lomolino 1998, Gaston 2003), but is rarely discussed as a control on unionoid distribution or abundance. Ecologists concerned with the conservation of *Margaritifera margaritifera* have mentioned the possibility of negative effects as global warming restricts the distribution of its coldwater salmonid hosts (Hastie et al. 2003). Likewise, Golladay et al. (2004) noted that climate change, along with groundwater withdrawals for irrigation, may increase drought-caused mortality of unionoids in the American Southeast. In addition to warming, climate change may change the amount of water and sediment delivered to stream channels, thereby destabilizing stream channels and harming mussel populations (cf. Hastie et al. 2003). While it is possible that the influence of climate has been badly underestimated, it is also possible that the influence of climate is masked by the substantial variation in the character of freshwater habitats within a climatic zone, along with the high frequency and effectiveness of dispersal barriers.

HOW FREQUENT IS HABITAT LIMITATION IN NATURE?

Habitat probably limits the extent and abundance of unionoid populations in many circumstances. Unstable sediments (Figs. 18 and 19; Vannote and Minshall 1982, Young and Williams 1983, Layzer and Madison 1995, Strayer 1999a, Johnson and Brown 2000, Hastie et al. 2001, Howard and Cuffey 2003), inadequate oxygen or excessive unionized ammonia (Buddensiek et al. 1993, Aldridge and McIvor 2003, Augspurger et al. 2003, Newton 2003, Mummert et al. 2003), anthropogenic pollutants (Fuller 1974, Goudreau et al. 1993, Naimo 1995, Augspurger et al. 2003), low temperatures (Hruška 1992, Heinricher and Layzer 1999, Beaty and Neves 2004), and dewatering (Miller and Payne 1998, Gagnon et al. 2004, Golladay et al. 2004) all commonly limit unionoid populations. In addition, sediments too soft or too hard for burrowing (Hastie et al. 2003) and currents too high for juvenile settlement (Layzer and Madison 1995, Payne and Miller 2000) may be important limiting factors. Thus, it would seem impossible to build a satisfactory model of unionoid distribution and abundance without considering habitat quality.

DENSITY-DEPENDENT FEEDBACKS ONTO HABITAT QUALITY

Mussels may change the quality or quantity of habitat, and thereby provide a mechanism for the regulation of their own population. Mussel populations may occasionally be so dense that physical space becomes limiting (Figs 26, 27). Juveniles settling into such dense mussel populations may be unable to find suitable sediment in which to burrow (or may even be consumed by adults as they settle). Nevertheless, as shown in Fig. 27, it takes a very dense mussel population to occupy even 10% of the bottom, so space limitation per se is not likely to limit mussel populations very often.

Mussel activities may also change habitat quality. Although it has been shown that dense populations of zebra mussels may appreciably deplete dissolved oxygen and raise ammonia concentrations (Effler and Siegfried 1994, Effler et al. 1998, Caraco et al. 2000), it seems unlikely that unionoids often reach sufficient densities to substantially affect oxygen and

FIGURE 26. A dense bed of the margaritiferid *Cumberlandia monodonta*. Photograph by W.N. Roston.

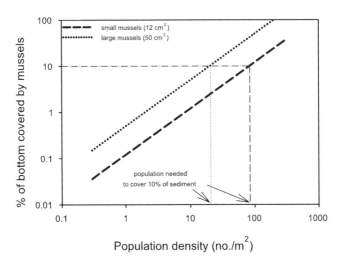

FIGURE 27. Percentage of the bottom covered by mussels of different sizes and population densities. Vertical and horizontal lines indicate the population densities required to cover 10% of the bottom. Compare to the population densities given in Fig. 42.

ammonia in the water column. Nevertheless, it seems possible that heavy biodeposition by dense mussel populations might have important effects on interstitial conditions, thereby affecting juvenile unionoids. Feeding mussels release large amounts of organic matter as biodeposits (feces and pseudofeces) (Vaughn et al. 2004). If the current is slow enough and the sediments are rough enough, much of this material may be deposited locally, in the mussel beds. Such material could be valuable food to juveniles, thereby providing a positive feedback between adult density and juvenile growth and survival. On the other hand, the decomposition of biodeposits could reduce oxygen concentrations and increase ammonia concentrations within stream sediments, reducing juvenile survival and growth. Recent lab studies (Zimmerman and de Szalay 2007) suggest that dense populations of adult unionids increase sediment compaction, which may also reduce the delivery of oxygen to interstitial juveniles (cf. Geist and Auerswald 2007). None of these possibilities has been investigated.

Finally, the presence of mussels may stabilize sediments and thereby improve habitat suitability (cf. Johnson and Brown 2000, Zimmerman and de Szalay 2007). Both the mass of the mussel shell and the burrowing behavior of a mussel might make the sediment less likely to erode during floods. This possibility has not received much attention, although initial lab studies (Zimmerman and de Szalay 2007) were inconclusive.

It is impossible to say at this point whether regulation of habitat quality provides an important feedback to mussel populations. It appears that such feedbacks may be important only in very dense mussel beds, and so may not be important in most natural mussel populations.

FIVE

HOSTS

The larvae of most unionoids are parasitic on fish, so it is reasonable to suspect that the distribution and abundance of host fish might be an important factor in limiting unionoid populations. Before discussing the use of fish by unionoids, though, it is worth briefly considering the exceptions to the basic life cycle, which may be of some use in understanding the nature of limitation by hosts. First, there have been occasional reports that some unionoids can dispense altogether with the fish host, and develop directly from larvae to juveniles. In the early 20th century, it was reported that *Obliquaria reflexa*, *Strophitus undulatus*, and *Utterbackia imbecillis* could develop without parasitism (Lefevre and Curtis 1911, Howard 1914). No one has repeated these observations, and all three of these species are now known to develop normally, using fish hosts (e.g., Tucker 1937, van Snik Gray 2002). It is unknown whether the early observations were erroneous, or whether the parasitic phase might be facultative. More recently, it has been shown definitely that *Lasmigona subviridis* is released from the parent as a juvenile, without a parasitic phase (Barfield and Watters 1998, Lellis and King 1998, Corey 2003). No one has yet described a parasitic glochidium in this species, although it seems too early to rule out the existence of a facultative parasitic phase. It is perhaps surprising that this species does not have an obviously distinctive pattern or distribution and abundance that can be attributed to the absence of dispersal and population limitation by hosts. Other than occasional observations that this species sometimes forms locally dense aggregations (e.g., Bailey 1891), which seems consistent with its direct development, its distribution and abundance seem to have attracted little notice. Nevertheless, species that can develop without a fish host may offer an interesting and useful contrast with other unionoids to gain insight into the importance of the parasitic phase in determining the abundance, distribution, dispersal, and genetic structure of mussel populations.

Second, a few unionoid species use amphibians for hosts. *Simpsonaias ambigua* uses the mudpuppy (*Necturus maculosus*) as its normal host, and apparently is unable to use fish for hosts (Barnhart et al. 1998). *Simpsonaias ambigua* is a widely distributed but rare species often said to occur in dense patches beneath large rocks (Call 1900, Howard 1951), probably as a result of the behavior and distribution of its host. Other species of unionoids can transform on amphibians as well, although fish are their normal hosts (Watters 1997, Watters and O'Dee 1998). It is not known how often unionoids parasitize amphibians in nature, and whether this might affect their dispersal and geographic distribution.

Returning now to the normal unionoid life history, we find that patterns of fish-host use are highly nonrandom, although we are far from having a complete catalog of mussel-host relationships. Some groups of fish are used heavily as hosts, while others are lightly used (Fig. 28). In some cases, this differential use seems easy to understand: darters (Percidae), sculpins (Cottidae), and nesting sunfishes (Centrarchidae) all are closely associated with the sediments, where they might contact mussels, whereas Clupeidae, which frequent the open water; Salmonidae, which live in cold waters; and Amblyopsidae, which live in caves, all would seem to make poorer hosts. Nevertheless, some of the patterns in Fig. 28 surprised me, particularly the infrequent use of suckers (Catostomidae) and lampreys (Petromyzontidae), which are widespread and common benthic fishes. Further, related mussel species tend to use related fish species as hosts (Table 6).

The number of known hosts varies widely across mussel species (Fig. 29). Most mussel species have 2–20 known host species; this figure underestimates actual host use, because host relationships are still imperfectly known. A few species have just a single known host, and appear to be genuinely specialized. Perhaps the most extreme example of such host specialization is the margaritiferid *Cumberlandia monodonta*; despite trials

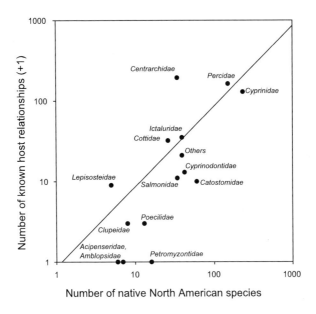

FIGURE 28. Patterns of host use in the North American unionoid fauna. The x-axis shows the number of native North American fish species in each family (from Jeschke and Strayer 2005, compiled from several sources), and the y-axis shows the number of times fish species from that family are listed as known hosts for mussel species. Specifically, these are relationships confirmed from natural or laboratory transformations, from Cummings and Watters (2005). The line shows the average for the entire fauna; families above the line (e.g., centrarchids) support more known host relationships than average, whereas families below the line (e.g., catostomids) support fewer than average. Note that both axes are logarithmic.

with 49 fish species from 28 genera and 12 families (and a couple of amphibians), no suitable host has yet been found (Knudsen and Hove 1997, Lee and Hove 1998, Baird 2000). At the other end of the spectrum are extreme host generalists such as *Strophitus undulatus* (36 known host species from 7 families) and *Pyganodon grandis* (30 known host species from 8 families), which transform on most fish species that are offered to them. Hyriids also appear to be host generalists (Walker et al. 2001). Bauer (1994) noted that anodontines, which have large glochidia, tend to use more host species than other unionoids. He suggested that species with large glochidia can transform in a shorter time on the fish, and therefore evade the immune

TABLE 6 *Related Mussel Species Use of Related Fish Species as Hosts, as Exemplified by Studies of* Pleurobema, Hamiota, *and* Epioblasma

Mussel species	Percidae	Cottidae	Cyprinidae	Centrarchidae	Other families
Pleurobema furvum	0	nd	**57%**	0	33%
Pleurobema clava	33%	nd	**67%**	0	0
Pleurobema decisum	0	nd	**12%**	0	0
Hamiota altilis	0	0	0	**50%**	0
Hamiota perovalis	0	nd	0	**43%**	0
Hamiota subangulata	0	nd	0	**75%**	50%
Epioblasma torulosa rangiana	**29%**	nd	0	0	17%
Epioblasma triquetra	17%	**100%**	0	0	0
Epioblasma brevidens	73%	**100%**	0	0	0
Epioblasma capsaeformis	38%	**100%**	0	0	0
Epioblasma florentina walkeri	**100%**	**100%**	0	0	0

NOTE: Data from Yeager and Saylor (1995), Haag and Warren (1997, 2003), Watters and O'Dee (1997), Haag et al. (1999), O'Brien and Brim Box (1999), Hove et al. (2000, 2003), O'Dee and Watters (2000), Rogers et al. (2001), Layzer et al. (2002), and Watters et al. (2005). The columns of the table are the percent of tested species in each fish family that served as hosts in laboratory trials. nd = no data. Dominant host families are given in boldface.

system of more fish species. Nevertheless, there is a wide range in host breadth within both the anodontines and other unionoids (Fig. 29).

One might expect that host specialists would be rarer and more vulnerable to human impacts than host generalists. There is a weak but interesting relationship between the breadth of host use and the conservation status of North American unionoid species (Fig. 30). All known host generalists are widespread and common, but host specialists range from being critically imperiled to being widespread and common.

Finally, it is important to note that the various hosts of a mussel species are not all equally important in supporting mussel recruitment. Many laboratory studies show large differences in transformation success on different host species (Table 7). Indeed, some recent studies suggest that there even may be important intraspecific variation in host relationships, with

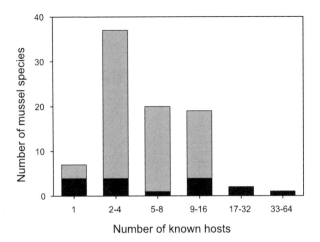

FIGURE 29. Number of known host species for each of 86 species of North American unionoids. Mussel species in the Anodontini are shown in black; all other species are shown in gray. Data include only mussel species that have been the subject of a life history study since 1970, and only native or naturalized fish species. From Cummings and Watters (2005).

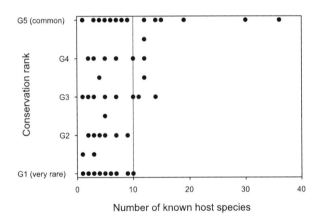

FIGURE 30. Relationship between the number of known host species (from Cummings and Watters 2005, estimated as in Fig. 28) and the conservation status of North American freshwater mussels (i.e., the "G" ranks of Nature Serve (2005).

TABLE 7 *Mean Number of Juveniles of* Hamiota subangulata *Produced by Several Species of Host Fish in Laboratory Trials*

Fish species (N)	% of fish producing juveniles	Number of juveniles/fish
Micropterus salmoides (21)	100	40.8
Micropterus punctulatus (11)	100	18.1
Gambusia holbrooki (8)	12.5	2.9
Lepomis macrochirus (10)	10	0.1

NOTE: From O'Brien and Brim Box 1999.

mussels better able to use populations of fish with which they co-occur than fish of the same species from other basins (Bauer 1987c, Rogers et al. 2001, Wächtler et al. 2001; but see Bigham 2002). Just as important, the behavior, seasonal movements, local distributions, and abundance of different host species will influence the actual exposure of each host to mussel glochidia and therefore its effectiveness as a host under field conditions. Thus, Martel and Lauzon-Guay (2005) showed marked differences in glochidial use of different hosts by the host generalist *Anodonta kennerlyi* in British Columbia lakes (Table 8). Studies like Martel and Lauzon-Guay's will be necessary to move from the important laboratory studies of host use to an understanding of actual use of various hosts in nature.

Once on the fish, glochidial mortality typically is high (>50%) even on suitable host fish (see Table 11.2 of Jansen et al. 2001, Hastie and Young 2001). Mortality may be density-dependent, although the strength and even the sign of density-dependence isn't clear (Bauer 1987c). A potentially crucial but poorly understood aspect of the mussel-host relationship is that even appropriate hosts develop resistance to glochidial infections after repeated exposure. This phenomenon was described in the early 20th century (Reuling 1919, Arey 1921, 1923, 1932), but has not received much attention. Thus, the transformation success of a mussel species on an individual fish falls after that fish has been repeatedly infected (Fig. 31, Bauer and Vogel 1987, Rogers and Dimock 2003, Dodd et al. 2005). It may take several infections for this resistance to develop fully (Rogers and Dimock 2003, Dodd et al. 2005), and it fades relatively quickly (Bauer and Vogel 1987, Dodd et al. 2006). The degree of resistance may vary across mussel-host systems—glochidial survival on resistant fish ranges from 28–100%

TABLE 8 *Density of Glochidia of* Anodonta kennerlyi *on Four Host Species in Three Lakes of British Columbia*

Lake	Cottus asper	Gasterosteus aculeatus	Salvelinus malma	Oncorhynchus clarki
Frederick	0.72	0.36	0.007	nd
Pachena	0.29	0.06	0.001	nd
Sarita	0.29	0.97	0.004	0.06

NOTE: Density is calculated by the number of encysted glochidia per cm^2 of head and fins. Data from Martel and Lauzon-Guay 2005. Note that the prickly sculpin (*Cottus asper*) is the preferred host in two of the lakes, but not in the third.

of survival on immunologically naïve fish (Bauer and Vogel 1987, Rogers and Dimock 2003, Dodd et al. 2005)—but appears never to completely prevent glochidial transformation. In addition, laboratory transformation success appears to be lower on older and larger fish than on young, small fish, even if none of these fish has had prior exposure to glochidia (Bauer 1987b).

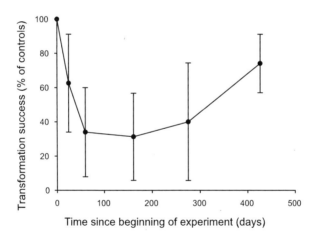

FIGURE 31. Development and subsequent loss of immunity (mean ± SD) in largemouth bass following repeated infections by glochidia of *Lampsilis reeveiana brevicula*. All fish received the first three infections (to provoke the immune response), then one of the following infections (to observe the loss of the immune response). Note the progressive loss of immunity between days 161 and 426, as well as the substantial variation among fish. From Dodd et al. (2006).

Very importantly, in some cases infected fish develop cross-resistance to glochidia of related mussel species (Fig. 32) (Dodd et al. 2005; but see Bauer 1991b, who found no evidence of cross-resistance). The degree of resistance may depend on the relatedness of the mussel species that are involved, and is relatively modest in the few cases where it has been demonstrated (Dodd et al. 2005).

Within a known host, both infestation rates and the number of glochidia per infested fish in nature often are lower on old, large fish than on young fish (Fig. 33; Tedla and Fernando 1969, Young and Williams 1984a, Bauer 1987b, Hastie and Young 2001), consistent with laboratory findings that older fish are intrinsically less suitable for glochidial transformation, and with the possibility that older fish may have become resistant from earlier glochidial infections. This pattern is not universal (e.g., Jokela et al. 1991, Blaĭek and Gelnar 2006), and the degree to which immunological resistance is responsible for this pattern is not clear.

The subject of immunity is critically important for two important questions about unionoid populations. First, immunity may cause intraspecific

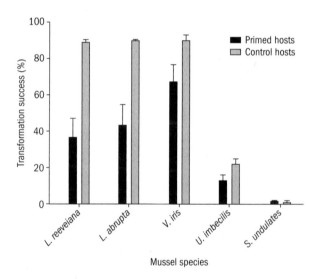

FIGURE 32. Reduction in transformation success (% of glochidia surviving until metamorphosis) with the development of acquired immunity. Gray bars show success on naïve hosts, while black bars show success on fish that had four or five previous infections by glochidia of *Lampsilis reeveiana brevicula*. From Dodd et al. (2005). Reprinted with permission from the Journal of Parasitology.

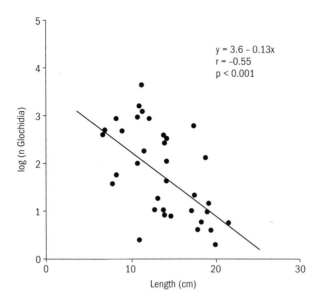

FIGURE 33. Number of glochidia of *Margaritifera margaritifera* carried by brown trout of different sizes in a small German brook, from Bauer (1987b).

competition for hosts, leading to density-dependent feedback between mussels and fish that may control mussel populations. Second, the existence of cross-species immunity raises the possibility of interspecific competition for hosts. We do not yet know if immunity in nature is strong enough to lead to significant host competition and ultimately affect the distribution and abundance of unionoids.

Are host infestation rates in nature high enough to induce significant immunological resistance to glochidial transformation? Year-long surveys of mixed-species fish communities often report glochidial prevalence rates of <10% (e.g., Weir 1977, Neves and Widlak 1988, Weiss and Layzer 1995), perhaps suggesting that host resistance would be unimportant. However, when studies of glochidial prevalence are restricted to known hosts, peak prevalence rates (probably the best measure of how many fish might develop immunological resistance) often are 50-100% (Table 9). Of course, such studies often are done around dense mussel populations, so the data in Table 9 probably represent the upper end of glochidial prevalence rates in nature. Nevertheless, it is clear that glochidial prevalence rates can be

TABLE 9 Peak Glochidial Prevalence on Known Hosts in Nature

Mussel species	Host species	Site	Peak prevalence	Source
Margaritiferidae				
Margaritifera margaritifera	Salmo trutta (0+)	Stac Burn, Scotland	77–83%	Young and Williams (1984a)
Margaritifera margaritifera	Salmo trutta (1+)	Stac Burn, Scotland	20–57%	Young and Williams (1984a)
Margaritifera margaritifera	Salmo trutta (0+)	Small German brook	100%	Bauer (1987b)
Margaritifera margaritifera	Salmo trutta (1+)	Small German brook	50%	Bauer (1987b)
Margaritifera margaritifera	Salmo trutta (2+)	Small German brook	13%	Bauer (1987b)
Margaritifera margaritifera	Salmo salar	South River, Nova Scotia	89–100%	Cunjak and McGladdery (1991)
Margaritifera margaritifera	Salmo salar (0+)	Five Scottish rivers	69–100%	Hastie and Young (2001)
Margaritifera margaritifera	Salmo salar (1+)	Five Scottish rivers	20–100%	Hastie and Young (2001)
Anodontini				
Alasmidonta heterodon	Etheostoma olmstedi	Mill River, MA	0–52%	McLain and Ross (2005)
Alasmidonta viridis	Cottus carolinae	Big Moccasin Creek, VA	11%	Zale and Neves (1982b)
Anodonta cygnea	Gasterosteus aculeatus	Shoulder of Mutton Pond, UK	100%	Dartnall and Walkey (1979)
Anodonta piscinalis	Perca flavescens	Lake Kuivasjärvi, Finland	95%	Jokela et al. (1991)
Anodonta piscinalis	Rutilus rutilus	Lake Kuivasjärvi, Finland	68%	Jokela et al. (1991)
Pyganodon grandis	Perca flavescens	Narrow Lake, Alberta	100%	Jansen (1991)

(continued)

TABLE 9 Peak Glochidial Prevalence on Known Hosts in Nature *(continued)*

Mussel species	Host species	Site	Peak prevalence	Source
Lampsilis radiata	*Perca flavescens*	Bay of Quinte, Lake Ontario, Ontario	87%	Tedla and Fernando (1969)
Lampsilis siliquoidea	Seven centrarchids and percids	Shannon Lake, MN	4–25%	Trdan (1981)
Medionidus conradicus	*Etheostoma flabellare*	Big Moccasin Creek, VA	100%	Zale and Neves (1982a)
Pleurobema oviforme	Six cyprinids	Big Moccasin Creek, VA	11–46%[a]	Weaver et al. (1991)
Venustaconcha ellipsiformis	*Etheostoma caerulum*	James River, MO	64%	Riusech and Barnhart (2000)
Venustaconcha ellipsiformis	*Etheostoma spectabile*	Spring River, MO	37%	Riusech and Barnhart (2000)
Villosa nebulosa	*Ambloplites rupestris*	Big Moccasin Creek, VA	84%	Zale and Neves (1982a)
Villosa vanuxemi	*Cottus carolinae*	Big Moccasin Creek, VA	88%	Zale and Neves (1982a)

Other Unionidae

[a] May include a few glochidia of other mussel species.

high enough to potentially induce significant resistance in host populations. Table 9 also shows a great deal of variation in glochidial prevalence rates, which may be important to mussel-host relations. As already noted, glochidia may occur less frequently on older, larger fish, at least in *Margaritifera*. There is no obvious pattern in Table 9 with respect to phylogeny or host breadth, although the data set is still small. Variation presumably also arises from differences in mussel density, water depth and habitat structure (leading to differences in probability of contact between host and glochidia), and fish abundance. Whatever the cause, these data suggest that we can expect to see the full range of variation in immunological status among wild host populations, from host populations that consist entirely of immunologically naïve fish to populations in which all older fish have at least some induced resistance to glochidial infections.

SIMPLE EPIDEMIOLOGICAL MODELS OF MUSSEL-FISH DYNAMICS

Simple epidemiological models (based on those presented in Vandermeer and Goldberg 2003) may be helpful in illuminating the potential roles of immunological resistance and competition for hosts. Define the following terms:

j = mussel survivorship from dropping off the fish until sexual maturity

a = probability that a glochidium will attach to a fish (these models assume that fish-glochidia encounters are independent events, and would need to be modified in cases in which fish-mussel contacts aren't independent; e.g., conglutinates, displaying females)

g = number of glochidia released per adult (i.e., number of glochidia per female/2) (in these simple models, g is modeled as density-independent, although it may be desirable to construct models with an Allee effect by modeling $g = f(N)$)

N = number of adult mussels

F = number of adult fish ($F = R + I + S$, assumed to be constant)

R = number of resistant (immune) fish

S = number of susceptible fish
I = number of infected fish
m_m = mortality rate of adult mussels
m_f = mortality rate of fish

Assume the following:

- F is constant (i.e., the size of the mussel population doesn't affect the size of the fish population)
- Fish remain immune forever (this assumption will be relaxed in later models)
- The distribution of numbers of glochidia on fish follows a Poisson distribution (no pun intended) with a mean and variance of gaN

The dynamics of this system are described as follows:

$$\frac{dN}{dt} = gajNS - m_m N.$$

The first term on the right-hand side of the equation is recruitment to the pool of adult mussels, and the second term is mortality. Next, in

$$\frac{dS}{dt} = m_f F - S(1 - e^{-gaN}) - m_f S,$$

the terms on the right-hand side of the equation are, in order, new recruits to the fish population, (which must equal the number of fish leaving via mortality because the total number of fish is constant); transfer of susceptible fish to the resistant pool as they are infected by glochidia; and mortality of susceptible fish. In

$$\frac{dR}{dt} = S(1 - e^{-gaN}) - m_f R,$$

the terms on the right-hand side of the equation are transfer of fish from the susceptible pool as they are infected by glochidia, and mortality of resistant fish, respectively.

When this system is at equilibrium, which is to say that

$$\frac{dN}{dt} = \frac{dS}{dt} = \frac{dR}{dt} = 0$$

then

$$\hat{S} = \frac{m_m}{gaj},$$

the proportion of the fish population that is susceptible to glochidial infection is

$$\frac{m_f}{(1 + m_f - e^{-gaN})},$$

and

$$\hat{N} = \frac{-\ln\left(1 - \frac{m_f(gajF - m_m)}{m_m}\right)}{ga}.$$

Thus, the number of susceptible fish in the population rises with the mortality rate of adult mussels, and falls as fecundity, the probability of glochidia-fish encounters, and survival of juvenile mussels increases. The proportion of the fish population susceptible to glochidial infection rises with the mortality rate of the fish population, falls with increasing probability of glochidia-fish encounters, and falls with increasing size or fecundity of the mussel population (Fig. 34). This model suggests that mussels may deplete the pool of suitable fish hosts, and that the proportion of the fish population that is susceptible is controlled chiefly by the turnover rate of the fish population, except at low loads of attached glochidia (field data on glochidial loads show that they often exceed 5/fish: Tedla and Fernando 1969, Young and Williams 1984a, Bauer 1987b, Jansen 1991, Jokela et al. 1991, Martel and Lauzon-Guay 2005). The size of the adult mussel population will rise with the number of fish (Fig. 35), with the mortality rate of fish (which supplies new, susceptible hosts), and with increasing survivorship of juvenile mussels. The population will fall with increasing mortality of adult mussels. None of these results is surprising.

Note that this result implies that sites with higher primary production or nutrient loads, which support higher fish production (Downing and Plante 1993, Randall et al. 1995), will support larger mussel populations

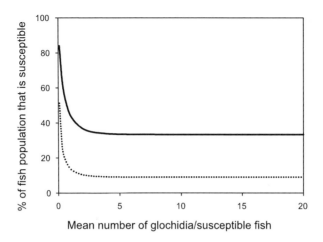

FIGURE 34. Model results for the percentage of a host fish population that is susceptible to glochidial infection as a function of mean glochidial load and mortality of the fish population (solid line = 0.5, dotted line = 0.1).

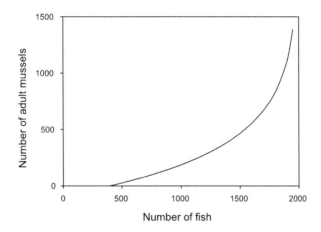

FIGURE 35. Example of model results showing how the number of adult mussels rises with the number of fish (F), for a particular combination of life history parameters; i.e., $a = 10^{-7}$, $g = 25,000$/adult, $j = 0.05$, $m_f = 0.25$, and $m_m = 0.05$. The mussel population is not viable at fish populations less than 400, and grows without bound when the number of fish is 2000.

(this result is independent of any food-limitation of the mussels themselves). Further, unionoid species that use long-lived fishes as hosts will have smaller populations than those that use short-lived hosts, all else being equal. Finally, mobile fish species (which will have a higher value of m_f) may support higher mussel populations, although the lowering of j by dropping juvenile mussels in inappropriate habitats may counteract this effect.

Finally, we can squeeze just a little more information out of these results by noting that the expression for $\hat{N}$ will be negative (meaning that the mussel population isn't viable) when

$$\ln\left(1 - \frac{m_f(gaj - m_m)}{m_m}\right) > 0,$$

which will occur if and only if

$$\left(1 - \frac{m_f(gaj - m_m)}{m_m}\right) > 1.$$

Likewise, the expression for $\hat{N}$ will be undefined (because the logarithm of negative numbers is undefined) if

$$\left(1 - \frac{m_f(gaj - m_m)}{m_m}\right) \leq 0.$$

Combining and simplifying these last two equations, and making the reasonable assumption that $m_f > m_m$, we find that

$$0 < (gajF - m_m) < 1,$$

which greatly constrains the range of possible life history parameters and fish populations that can sustain viable mussel populations.

This simple model is highly unrealistic, particularly in its assumption that fish become completely immune to glochidial infection after just one exposure. However, such simple analytical models rapidly become intractable (at least for me!) when more complexity is introduced, so I will use simulation models to explore the sensitivity of the fish-mussel system to changes in the strength of the fish immune response. The system is conceived just as described above, except that fish mortality and the strength of the mussel immune response are allowed to vary, and specific values are chosen for other parameters as follows: $j = 0.05$ or 0.025 (cf. Young and Williams 1984a); $a = 10^{-7}$ (I estimated this value by assuming 25 glochidia per susceptible fish at an adult mussel population of 10,000; i.e., 10 mussels/fish, which equals gaN under the model assumption of Poisson-distributed glochidia)); $g = 25,000$/year (cf. Haag and Warren 2003); $F = 1000$; and $m_m = 0.05$/year. Resistance to glochidial infection is graded from 0.2 (a resistant fish will carry 80% as many glochidia to transformation as a naïve fish) to 1 (a resistant fish won't carry any glochidia through to transformation, as in the earlier model).

Both the size of the mussel population and the degree to which mussels deplete the pool of susceptible hosts depend on the strength of resistance that fish develop (Fig. 36). The mortality rate of the fish population (indicating the supply rate of immunologically naïve fish) also is important, as in the previous model assuming complete immunity. Figure 36 also shows that there are large regions of parameter space in which host populations are not limiting (mussel populations grow without bound). In these regions, factors other than host availability must control mussel populations.

These models allow mussel populations to grow without bounds because they allow individual fish to carry an unlimited number of glochidia, obviously an unrealistic assumption. There must be a limit to the number of glochidia that can transform on an individual fish. Perhaps glochidial survival on a host is density-dependent: see Bauer and Vogel (1987) for an example of both density-dependence and (!) inverse density-dependence in glochidial survival. Such density-dependence can be added to the models just presented by adding a logistic term to express the proportion of glochidia that successfully transform. Specifically, I modified the previous model by assuming that the proportion of glochidia that transform successfully is

$$\left(\frac{(50 - I)}{50}\right)$$

where I is the mean number of glochidia that initially attach to the host. The number 50 was chosen to represent the number of glochidia/fish that might reasonably be expected to complete metamorphosis (see references cited in Table 9). (Note that this model is not quite technically correct, because the term for density dependence assumes that all fish carry the same number of glochidia. Technically, the degree of density-dependence should vary from fish to fish depending on the actual glochidial loads, which I've previously assumed were Poisson-distributed. However, the simplified

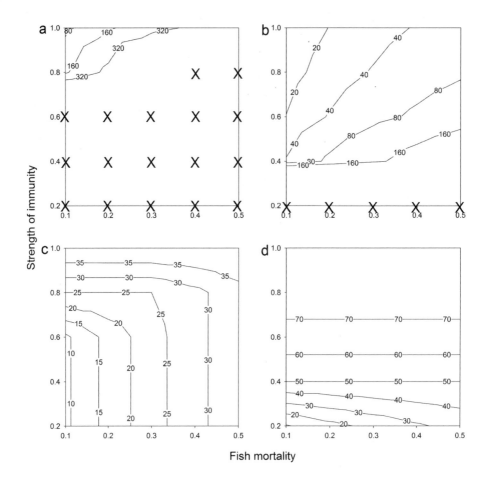

FIGURE 36. Results of simulation models of the effects of fish immunity and mortality on the size of mussel populations and the percentage of the fish population that is resistant to glochidial infections. Panels (a) and (b) show the equilibrial size of the mussel population; parameter values marked "X" gave unbounded population growth. Panels (c) and (d) show the proportion of the fish population that is resistant to glochidial infection. Panels (a) and (c) are based on a value of j (survivorship from the time that a mussel excysts from its host until it reproduces) of 0.05; panels (b) and (d) assume $j = 0.025$.

density-dependence term that I used is much easier to deal with and presumably produces qualitatively the same effect as a more technically accurate term).

The results of this model are shown in Fig. 37. As in the simpler model (Fig. 36), this model shows that the strength of the fish immune response is critically important, and that the mortality rate of the host population also is important. The density-dependent glochidial survival prevents unlimited population growth of the mussels, although a very wide range of equilibrial mussel population densities can occur, depending on the fish

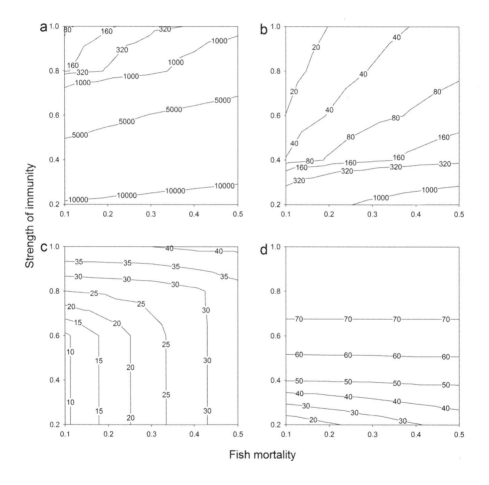

FIGURE 37. Same as Fig. 36, but with density-dependent survival of glochidia on fish hosts. See text for details.

immune response and mortality rate. Adding density-dependent glochidial survival to the model had very little effect on the fractionation of the fish population into susceptible and resistant individuals.

These models point out the potential importance of the immune responses of fish to the distribution and abundance of their mussel parasites. This is an area in which very little is known at present. Clearly, if we plan to build mechanistic models of mussel demography, we will need much more quantitative information on this key subject.

In addition, I think that further investigation of models of fish-glochidia interactions might be worthwhile. In particular, models using more realistic descriptions of fish immunity might be worth exploring. Multispecies models of fish-glochidial interactions with various degrees of cross-resistance also might shed light on the conditions for coexistence of diverse mussel communities (cf. Rashleigh and DeAngelis, 2007).

HOST LIMITATION IN NATURE

Several different lines of evidence suggest that unionoid populations may be limited by host availability in nature. Most directly, mussel populations have expanded or contracted following expansion or contraction of the ranges of their hosts. Thus, Smith (1985) noted that *Anodonta implicata* reappeared in a section of the Connecticut River above the Holyoke Dam after a fish "elevator" was provided to carry American shad (its presumed host) over the dam. Conversely, *Fusconaia ebena* apparently has not reproduced in the upper Mississippi River since Lock and Dam 19 blocked migrations of its host fish (the skipjack herring) in 1913 (Kelner and Sietman 2000), and will presumably disappear once the old animals recruited before dam construction die. These are examples of all-or-nothing limitation in which total absence of the host prevents a mussel species from living in otherwise suitable conditions.

Host abundance may also modulate mussel abundance. Both Haag and Warren (1998) and Mulcrone (2004) reported correlations between local densities of mussels and densities of their known hosts. Interestingly, Haag and Warren (1998) found that such correlations occurred only for host specialists that did not display to attract hosts. They suggested that mussel species that have many hosts or use displays to attract hosts are able to overcome host limitation. The relationship between breadth of host use and conservation status (Fig. 30) also suggests the importance of host availability. Finally, several authors have noted correlations between fish community structure and mussel community structure, which suggests that the fish community may be controlling the mussel community. For instance, species richness of both fishes and mussels rises with increasing stream size (Watters 1992). Of course, such correlations may indicate merely that both fishes and mussels are controlled by the same environmental factors. Nevertheless, the formal statistical analyses of Watters (1992) and Vaughn and Taylor (2000) suggest that these correlations arise at least in part because fish community structure controls mussel community structure.

It is clear, however, that mussel populations are not simply controlled by host availability. Several authors (e.g., Strayer 1983, Gordon and Layzer 1993, Bauer 1991a) have noted that the geographic ranges of mussel species often are much smaller than those of their hosts (Fig. 9). As already noted, local correlations between host abundance and mussel abundance seem to apply to only a subset of the mussel community (Haag and Warren 1998), although few such correlations have yet been attempted. Bauer (1991b) reported that the number of encysted glochidia in a series of populations of *Margaritifera margaritifera* was related to the number of adult mussels at the site, suggesting that host densities were of secondary importance. Finally, Geist et al. (2006) could find no correlation between the number of host fish and the occurrence of successful recruitment in European populations of *Margaritifera margaritifera*.

As with the other factors that might control unionoid populations, we conclude that host availability surely controls unionoid distribution and abundance in some cases, but not in all cases, and that our present knowledge is inadequate to specify the frequency or severity of the limitation, or describe the conditions under which it is most likely to occur.

Humans have had large effects on freshwater fish populations through habitat modification, species introductions, and harvest (e.g., Fig. 45, Chapter 8; Allan et al. 2005). To the extent that mussel populations are host-limited, these changes may cascade down to affect the mussel community. As in the case of *Fusconaia ebena*, these cascading effects may take decades or even centuries to be fully expressed. Thus, if mussel populations are commonly host-limited, they may still be reacting to the profound changes that humans caused to freshwater fish communities through the 19th and 20th centuries.

We clearly need a better sense of how often, how severely, and under what conditions host availability limits mussel populations. It is important to know if mussels are able to deplete the pool of available hosts through immunological resistance to glochidial infection—what proportion of potential host populations are resistant to glochidial infection? Another possibly important source of density-dependence is the density-dependence in glochidial survival investigated inconclusively by Bauer and Vogel (1987). Is either form of density-dependence strong enough to have demographic effects? How high do mussel densities have to be, or how low do fish densities or turnover rates have to be, for density dependence to become important? The whole issue of cross-species resistance to glochidial infection has barely been investigated, but will determine the extent to which interspecific competition for hosts occurs. Can host specialists deal with immune systems better than host generalists? If so, there might be a tradeoff that would allow coexistence of multiple species, host specialists dominating in the high-quality and stable habitats, while host generalists with their higher dispersal and lower requirements for minimum host densities could dominate in marginal habitats. Finally, it would be useful to know more about infestation rates in nature. To be relevant, these rates should be conducted at local scales around mussel beds, to reflect the population of potential hosts that the mussels are actually exposed to. Available data suggest an extraordinarily wide range of infestation rates in nature, which at face value suggest that the strength of host limitation might likewise vary over a very wide range. Information on host use could also be helpful in understanding dispersal in mussel populations. For instance, information on spatial variation in infestation rates around well defined mussel beds might shed light on the spatial extent of dispersal of glochidia and fish hosts.

SIX

FOOD

The quality and quantity of food are often regarded as among the primary factors that regulate the distribution and abundance of organisms in nature. It is thus curious that little attention has been given to the possibility that food might limit the distribution or abundance of unionoid mussels. We do not even have a clear idea what unionoid food *is*, and know even less about the severity or extent of food limitation in pearly mussel populations.

WHAT DO MUSSELS EAT?

Information on unionoid diets has come from analyses of the gut contents of animals collected from the field, laboratory feeding trials, studies of mussel anatomy and biochemistry, laboratory studies of growth and survival of mussels fed different diets, tracer studies based on stable isotopes and fatty acids of mussels and their putative food sources, and inferences based on spatial and temporal dynamics of putative food resources. These diverse studies have not resulted in a clear consensus about the nature of the unionoid diet.

Classically, freshwater mussels were thought of as suspension feeders, subsisting chiefly on phytoplankton, zooplankton, and particulate detritus (e.g., Allen 1914, Coker et al. 1921, Churchill and Lewis 1924, McMahon and Bogan 2001). More recent work has raised the possibility that unionoid mussels may feed on suspended bacteria (Silverman et al. 1997, Nichols and Garling 2000), fungal spores (Bärlocher and Brendelberger 2004), dissolved organic matter (cf. work on zebra mussels by Roditi et al. (2000) and Baines et al. (2005)), or sedimented organic matter (Yeager et al. 1994, Raikow and Hamilton 2001, Nichols et al. 2005). Consequently, our understanding of unionoid diets is unresolved and in flux.

The earliest attempts to understand the unionoid diet were based on examinations of the gut contents of animals taken from the field. These studies reported that unionoid guts contained a heterogeneous mix of phytoplankton, small zooplankton (i.e., rotifers and small cladocerans), detritus, and inorganic particles (sand and silt) (Allen 1914, Churchill and Lewis 1924). More sophisticated recent work has reported much the same general picture, but has also raised the possibility that species living side by side may contain somewhat different material in their guts (Bisbee 1984, Parker et al. 1998, Nichols and Garling 2000, Vaughn and Hakenkamp 2001). In a general way, the particles in unionoid guts resemble those suspended in the seston (Coker et al. 1921, Nichols and Garling 2000). Taken together, field studies of gut contents suggest that unionoids are capable of capturing a wide range of particle types and sizes, and that interspecific differences in particle capture may exist. Nevertheless, examination of gut contents cannot identify which of the captured particles are actually assimilated or needed by the mussel; many particles pass through the digestive system without being assimilated (e.g., Coker et al. 1921, Nichols and Garling 2000). Further, gut analyses have not been useful in determining the nature of food that is not readily recognizable visually (detritus, bacteria, or dissolved matter).

Several laboratory studies have estimated the ability of unionoids to capture food. These studies have focused most often on clearance rates and size-selection by the mussels. Again, these studies have found that mussels can capture a wide range of particles, from just under 1 μm in size (e.g., large bacteria and the smallest phytoplankton) up to at least 40 μm (small zooplankton) (Brönmark and Malmqvist 1982, Paterson 1984, 1986, Vanderploeg et al. 1995). Not all particles in this broad size range are captured with equal efficiency (Fig. 38; Brönmark and Malmqvist 1982, Paterson 1984, 1986, Vanderploeg et al. 1995, Baker and Levinton 2003, Beck and Neves 2003).

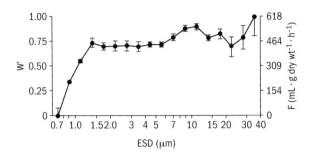

FIGURE 38. Filtration rate (F) and retention efficiency (W′) in *Lampsilis radiata siliquoidea* feeding on natural seston from Lake St. Clair, Michigan, as a function of the equivalent spherical diameter (ESD) of the particles. W′ is the ratio of particle clearance in any size category to particle clearance on the preferred size-class (F_i/F_{pref}). From Vanderploeg et al. (1995).

The ability to capture particles varies among species, life-stages, and sexes, although these differences have not been described in much detail. Thus, Silverman et al. (1997) found that stream-dwelling species were much better than pond-dwelling species at capturing bacteria, which agrees with observed morphological differences in the gills of these two groups of species. Beck and Neves (2003) found that particle selection differed markedly across age in *Villosa iris*. The filtration rates of gravid females are much lower than those of males or non-gravid females, apparently because use of the gills by gravid females to brood larvae compromises their feeding function (Tankersley and Dimock 1993).

Reported laboratory filtration rates vary widely (see, e.g., the summaries of Kryger and Riisgard 1988, Walker et al. 2001). Laboratory feeding studies have been done under a wide range of conditions, usually highly artificial, and it is likely that many of the results are unreliable (Riisgard 2001); in particular, many of the estimates of clearance rates probably are too low. An example of what is probably a reliable estimate is Kryger and Riisgard's (1988) estimate that the filtration rate of *Anodonta anatina* can be summarized as

$$FR = 26.4\ DM^{0.78}$$

where FR is the filtration rate in L/d (at 19–20 °C) and DM is the shell-free dry mass of the animal in grams. This suggests that a typical adult mussel of dry mass 0.5–2 g might remove the particles from ~15–45 L/d. As in the studies of gut contents, however, laboratory studies of particle capture identify merely what the animal is capable of capturing, not what it assimilates or requires.

Laboratory studies also have been valuable in elucidating mussel feeding behaviors. In particular, laboratory studies have shown that both juvenile (Yeager et al. 1994, Gatenby et al. 1996) and adult (Nichols et al. 2005) unionoids can take up benthic food via some form of deposit-feeding. Marine scientists have also used laboratory studies to examine the effects of environmental factors such as current speed on bivalve feeding (see summary by Wildish and Kristmansen 1997), but this line of inquiry has not been much pursued for unionoids.

Finally, laboratory feeding studies have provided insight into what foods are actually assimilated by unionoids and can support their growth and development. There is perhaps potential to use this approach more heavily to investigate the degree to which different foods (appropriately labeled) can be assimilated by unionoids. Nevertheless, it is clear that different sorts of food have vastly different abilities to support unionoids. Most significantly, algae high in polyunsaturated fatty acids seem to be essential for the growth and survival of juveniles (Gatenby et al. 1997). Even for adults, many apparently suitable foods do not support growth, reproduction, and survival (Nichols and Garling 2002). Recent laboratory studies showing that zebra mussels may meet a large part of their energy needs by taking up dissolved organic matter from the water (Roditi et al. 2000, Baines et al. 2005) suggest that unionoids likewise might be able to use this food resource.

Additional clues to the unionoid diet come from studies of mussel anatomy and biochemistry. Thus, Silverman et al. (1997) noted that the spacing of cirri on the gills was correlated with the ability of different species to capture small particles (bacteria), and Payne et al. (1995) noted that freshwater bivalves living in turbid waters had higher palp:gill area ratios than those from clear waters. An old study by Crosby and Reid (1971) showing that unionoids contain cellulase has important consequences for understanding the kinds of foods that these animals can use, but seems not to have been widely appreciated. There can be large differences in cellolytic activity among species, life-stages, and sites (Johnson et al. 1998, Areekijseree et al. 2006), but it is not yet clear how these differences might affect growth or demography, or contribute to niche separation among species.

Recent advances in biochemistry and especially in stable isotopes make it possible to infer useful information about unionoid diets from the chemical and isotopic composition of field-collected animals. So far, this work has been limited and has yielded ambiguous results. Bunn and Boon (1993) found that the Australian hyriid *Velesunio ambiguus* contained less ^{13}C than any potential food source they analyzed, and suggested that the mussel might depend on a mixture of methanotrophic bacteria and other items (plant detritus, plankton). Nichols and Garling (2000) found that unionoids contained vitamin B_{12}, which must have been obtained from bacteria, and concluded that the ^{13}C and ^{15}N content of these animals suggested that they were supported chiefly by bacteria. Christian et al. (2004) also concluded that bacteria formed a major part of the unionoid diet in small Ohio streams. These results are problematic, though, because in all studies the ^{13}C content of unionoids was lower than that of any food source analyzed, suggesting that additional, highly ^{13}C-depleted foods (e.g., methanotrophic bacteria) had been consumed by the mussels, or that the mussels were selectively assimilating ^{13}C-depleted fractions from a heterogeneous fine particulate organic matter pool. Raikow and Hamilton (2001) found that the ^{15}N content of unionoids in an experimentally labeled Michigan stream suggested that they were predominately deposit-feeding, but they could not rule out the possibility that the mussels were differentially assimilating different fractions of ingested particles.

Generally, stable isotope studies of unionoid diets have suffered from indetermination and inadequate resolution of mixed pools (e.g., "suspended particles," which includes various algae, fungi, bacteria, and detrital particles, each with its own isotopic signature). If a simple mixing model is used (the usual practice in stable isotope studies), a study using N isotopes can unambiguously estimate the dietary contribution of no more than $N+1$ food sources. If the food sources overlap in isotopic content or are composed of isotopically heterogeneous mixtures, as is often the case, then even this discrimination may not be possible. All existing stable isotope studies of unionoid diets have tried to estimate the contribution of multiple, often heterogeneous, food sources, from a small number of often poorly resolved isotopes, so the authors were not able to make unambiguous estimates of the contributions of these sources. These problems will plague future attempts to use stable isotopes to identify unionoid diets, and can be solved only by bringing additional information (e.g., additional tracers, more constrained food web models) to bear on the problem.

Finally, in a few cases it may be possible to gather useful information about unionoid diets by observing the impacts that mussels have on food resources in the field. Thus, Welker and Walz (1998) observed steep declines in rotifer abundance as the River Spree flowed over dense ($14/m^2$) unionoid beds, even though the rotifers' reproductive parameters did not change, suggesting that unionoids do ingest large numbers of these small zooplankton.

To sum up, we know that unionoids are able to capture a wide range of suspended particles that might serve as food: large bacteria, phytoplankton, small zooplankton, organic detritus, and perhaps dissolved organic matter. Different species of unionoids—indeed, different species of freshwater bivalves and other suspension-feeders—overlap broadly in the kinds of particles they capture and could potentially compete for food. We know that the foods that unionoids are able to capture are of very unequal value to them: many captured items are not even assimilated, while others appear to be essential to survival, growth, and reproduction. Juvenile and adult unionoids can take up sedimented food particles as well, although feeding rates, particle selection and the quantitative significance of deposit-feeding in the unionoid diet in nature are not known. Work on marine bivalves strongly suggests that unionoid feeding abilities and rates probably vary with environmental conditions. There are some suggestions that unionoids might be able to adjust their abilities to capture and process particles under different environmental conditions, a valuable (and expected) trait for animals that are routinely exposed to both widely varying food types and concentrations and widely varying environmental conditions. Despite this considerable insight into the unionoid diet, we are far from being able to provide a quantitative measure of food availability to unionoids from a water or sediment sample.

FOOD LIMITATION IN NATURE

I believe there are three overarching questions regarding the influence of food resources on unionoid populations: (1) how often and under what conditions are unionoid populations limited by inadequate food? (2) how often and under what conditions do unionoids themselves control ambient food resources (thereby raising the possibility of density-dependent population control via food resources)? and (3) has widespread anthropogenic eutrophication increased the size of unionoid populations by increasing food availability? At the start, it is important to distinguish between two kinds of food limitation: limitation of individual mussels and limitation of mussel populations. An individual mussel is food-limited if its growth, size, survival, fecundity, or fitness can be increased by giving it more or better food, or decreased by giving it less or worse food. Likewise, a mussel population is food-limited if its size, density, or extent can be increased by giving it more or better food, or decreased by giving it less or worse food. A population of food-limited mussels need not itself be food-limited. For example, if the size of a mussel population is strictly controlled by the number of suitable hosts, increasing food supply may increase growth, size, or fecundity of individual mussels without increasing population size. On the other hand, fecundity of mussels usually is a strong function of body size (Fig. 39; Byrne 1998, Haag and Staton 2003), so if well-fed mussels grow faster or larger, food-limitation of individual mussels may easily have demographic consequences. Thus, food limitation

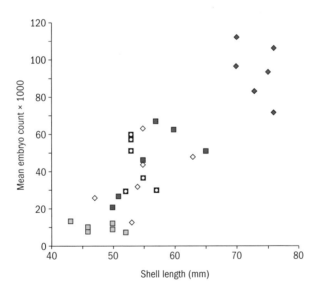

FIGURE 39. Fecundity (clutch size) of the hyriid mussel *Hyridella depressa* in six Australian sites (different symbols) as a function of shell length (Byrne 1998). With kind permission of Springer Science and Business Media.

of individual mussels is necessary but not sufficient to cause food limitation of mussel populations.

FREQUENCY AND SEVERITY OF FOOD LIMITATION IN NATURAL POPULATIONS OF MUSSELS

Given the difficulty in identifying or quantifying food resources for mussels, and the lack of attention paid to food limitation of mussels by ecologists, we are far from being able to offer a comprehensive assessment of how often and how severely mussel populations are limited by food. Nevertheless, we have evidence that food limitation may occur in nature, and can offer some suggestions about the conditions under which food limitation is most likely to occur.

The strongest evidence for food limitation of a mussel population probably comes from a study of the effects of the zebra mussel invasion on unionoids in the Hudson River (Fig. 40). A massive population of zebra mussels developed in the Hudson beginning in 1992, causing biomass of phytoplankton and small zooplankton to fall by 80–90% (although numbers of suspended bacteria rose) (Caraco et al. 1997, 2006, Findlay et al. 1998, Pace et al. 1998). At the same time, unionoid populations fell by 65–99.7% between 1992 and 1999, and body condition of unionids (i.e., body mass at a given shell size) fell by 20–30%. For some unknown reason, unionoids were not heavily infested by zebra mussels in the early years of the invasion of the Hudson, so the drastic changes in the Hudson's unionoids could not be attributed to fouling, as has been done elsewhere (e.g., Haag et al. 1993, Ricciardi et al. 1995, Schloesser et al. 1996). The most plausible explanation for the events in the Hudson is that zebra mussels severely reduced availability of food for unionoids, and that food-limitation increased mortality and decreased recruitment of the unionoids. Other reported declines in unionoid populations following the zebra mussel invasion (e.g., Ricciardi et al. 1995, Schloesser et al. 1996, Strayer 1999b) may also have been caused at least partly by food limitation, although usually attributed solely to fouling.

Other evidence for food limitation of mussel populations is weak. Mussels often are abundant in lake outlet streams and other areas with high phytoplankton biomass (Ostrovsky et al. 1993, Vaughn and Hakenkamp 2001), suggesting that these populations might be food-limited. Bauer (1991a) proposed that the relatively high metabolic rates of European unionids

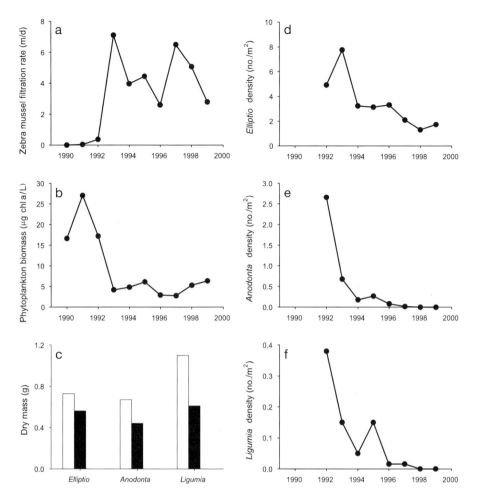

FIGURE 40. Response of plankton and unionoid populations in the early years of the zebra mussel invasion of the Hudson River, New York, suggesting food-limitation of the unionoids. a. riverwide filtration rate of the zebra mussel population during the summer; b. mean phytoplankton biomass during the growing season (May–September); c. average dry body mass of a 60-mm long animal of *Elliptio complanata*, *Anodonta implicata*, and *Ligumia ochracea* before (white bars) and after (black bars) the zebra mussel invasion; d. areally weighted density of *Elliptio complanata*; e. areally weighted density of *Anodonta implicata*, and f. areally weighted density of *Ligumia ochracea*. All differences between pre-invasion and post-invasion years are significant at p <0.01. From Strayer and Smith 1996, Caraco et al. 1997, Strayer et al. 1999, and Strayer et al., unpublished.

TABLE 10 *An Example of a Demographic Analysis of the Consequences of Increased Growth Rates and Reduced Life Span Induced by Eutrophication*

Age (x)	L_x, pre	L_x, post	l_x, pre	l_x, post	m_x, pre	m_x, post
1	17.7	24.3	1	1	0	0
2	28.5	44.6	1	1	0	0
3	42.1	60.6	1	1	0.10	0.29
4	51.0	74.5	1	1	0.17	0.54
5	55.1	77.7	1	0.5	0.22	0.62
6	58.5	84.5	0.6	0.29	0.26	0.79
7	63.5	100	0.5	0.04	0.34	1.31
8	72.2		0.25	0	0.49	
9	75.4		0.1		0.56	

NOTE: Based on data of Arter (1989) from Lake Hallwil, Switzerland. Pre-eutrophication data are represented by "soft" shells and living specimens from 1919; post-eutrophication data are represented by "hard" shells and living specimens taken in 1986. L_x, l_x, and m_x respectively are the shell length, survivorship and fecundity schedules at age x. Growth data from Arter (1989); I assumed that reproduction begins at age 3, fecundity is proportional to the cube of shell length (cf. Haag and Staton 2003), survivorship is represented by the age distribution of the animals collected, and the pre-eutrophication population had an intrinsic growth rate of 0 (i.e., was stable). Because of these untested assumptions, this analysis is intended to illustrate the procedure for analyzing such data, rather than being a true analysis of the eutrophication process in Lake Hallwil. This analysis gives a post-eutrophication intrinsic rate of increase of 0.08/year; thus, eutrophication increased population growth.

keep them out of food-poor habitats that *Margaritifera margaritifera*, with its low metabolic rate, is able to tolerate, suggesting that species-specific differences in food-limitation may determine unionoid distribution and abundance. Other evidence for food-limitation of mussel populations, either positive or negative, is lacking.

On the other hand, there is ample evidence that individual mussels may be food-limited. Good examples come from experiments in which mussels were transplanted from one site to another. For instance, Kesler et al. (2007) found growth rates of *Elliptio complanata* were rapid in Yawgoo and Tucker Ponds in Rhode Island, but essentially zero in nearby Worden Pond. Animals moved from Worden to Tucker or Yawgoo Ponds rapidly increased their growth rates, whereas animals moved in the opposite

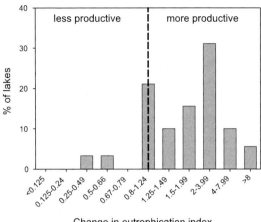

FIGURE 41. An example of the extent and severity of human-induced eutrophication. Change in an index of eutrophication (which includes information on nutrient loads and concentrations, peak phytoplankton biomass, and aquatic plant communities) between a "baseline" period in the 1930s and the early 1990s in a series of 90 British lakes. The change is calculated as the ratio of the eutrophication index in the 1990s to that in the 1930s; note that most lakes have become more than 50% more eutrophic during this period. Because many British lakes had already been eutrophied by the 1930s, this figure underestimates the severity of eutrophication in this region. Similar changes have been observed in many other parts of the world. From Moss et al. (1996).

direction stopped growing. Likewise, Walker et al. (2001) described an experimental translocation of the hyriid *Hyridella depressa* from an oligotrophic lake to stream sites above and below a sewage treatment plant (Walker et al. 2001). Animals moved to the enriched sites had higher growth and about double the fecundity of animals in less productive sites. Bauer (1998) found that only 5–54% of the female *Margaritifera margaritifera* in German populations reproduced in a given year, and that these reproductive animals contained a higher mass of non-reproductive tissue than non-reproductive females of the same shell length. He suggested that females reproduce only after they accumulate sufficient energy reserves. Arter (1989) found that growth rates of *Unio tumidus* rose after Lake Hallwil, Switzerland, was eutrophied by fertilizer and sewage. These examples suggest that food-limitation of individual mussels may be at least fairly widespread, satisfying at least some of the conditions for food-limitation of mussel populations.

It is worth noting that the higher growth rates induced by better food (or higher temperatures) typically are associated with shorter life spans (e.g., Arter 1989, Bauer 1991b). The overall effect of faster growth rates (earlier maturity and higher fecundity, but shorter life span) could be either to increase or decrease the population growth rate and size; a formal demographic analysis like that shown in Table 10 is required to judge the size and even the sign of the effect of eutrophication.

We do not know how often mussel populations are food-limited. Nevertheless, we can guess that food limitation will occur most often under three circumstances. First, if the environment is simply unproductive (e.g., nutrient-poor and with low allochthonous inputs), then there may not be enough food to support optimal survival, growth, and reproduction of mussels, which may in turn limit mussel populations. Such limitation by an unproductive environment was invoked by Bauer (1994) to explain the slow growth and intermittent reproduction of *Margaritifera margaritifera* in unproductive streams. It may be highly significant that human activities have increased nutrient inputs to many fresh waters, leading to large (i.e., several-fold) increases in phytoplankton biomass (e.g., Fig. 41; Heathwaite et al. 1996, Carpenter et al. 1998, Smith 2003, Smith et al. 2003). Likewise, dams have increased residence times and water clarity of many river systems (Rosenberg et al. 1997), which should lead to higher phytoplankton production. Because these changes are global in extent and of large magnitude (often more than twofold), they may have increased growth rates and population sizes of unionoids in many bodies of water, if these animals are food-limited. Such effects could have counterbalanced or even overridden any negative effects of eutrophication on unionoid populations (see Chapter 4, Habitat).

Second, food limitation may occur if consumers have high enough feeding rates to depress the quantity or quality of ambient food resources. I have already discussed the conditions under which suspension-feeders are most likely to be able to depress food resources (Strayer 1999b). Briefly, suspension-feeders can control food resources if their aggregate clearance rates are high relative to the renewal rates of the food supply (cf. Dame 1996). For a food resource whose growth is controlled by logistic growth, a grazer will reduce equilibrium food concentrations to

$$\hat{F} = K_f \left(\frac{r-g}{r}\right)$$

where $\hat{F}$ is the equilibrium food concentration, K_f is the carrying capacity of the food resource, r is its maximum intrinsic growth rate, and g is the grazing rate of the bivalve population. Thus, food resources are reduced by the size of the grazing rate, expressed as a fraction of the food's growth rate. In standing waters, where phytoplankton probably is the most important food, the aggregate clearance rates of suspension-feeders (as proportion of the mixed zone occupied by the population) may be compared to the growth rates of phytoplankton. Note that the relevant growth rate may be that of some nutritionally valuable fraction of the phytoplankton that grows much more slowly than the aggregate growth rate of all phytoplankton. Because the relevant measure is the volumetric grazing rate, deep water-columns will dilute the effects of mussel feeding. If grazing rates are much higher than growth rates, then suspension-feeders will greatly reduce phytoplankton biomass (and probably change its composition), whereas if phytoplankton growth rates are much higher than clearance rates, such control is unlikely. There are relatively few analyses of unionoid grazing rates in nature, but Fig. 42 gives an idea of what sorts of rates might be encountered.

In many cases, mussel grazing increases light penetration and nutrient concentrations (Arnott and Vanni 1996, Strayer et al. 1999, Vaughn and Hakenkamp 2001, Vaughn et al. 2004), leading to increased per capita phytoplankton growth. This feedback loop will lessen the ability of mussels to control phytoplankton biomass (Caraco et al. 1997, 2006) below that suggested by the simple logistic equation given above. The ability of suspension-feeders to locally deplete food may be substantially increased if the water column is imperfectly mixed, producing a food-depleted zone immediately around the mussel bed (MacIsaac et al. 1999, Ackerman et al. 2001, Edwards et al. 2005). Thus, unionoids are most likely to be food-limited as a result of their own activities in lakes if (1) their population density is high; (2) the water column is shallow; (3) phytoplankton growth rates are low; (4) the feedback between mussel grazing and increased phytoplankton growth is weak; and (5) the water column is imperfectly mixed.

The situation in running waters is more complicated because food may be transported from upstream, from the benthos, and from the banks

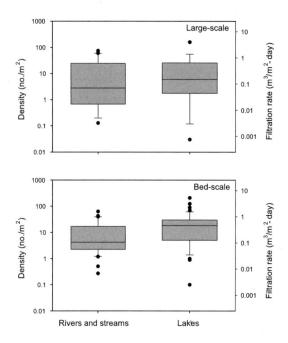

FIGURE 42. Reported densities of unionoids over large scales (entire lakes or long stream reaches) (upper panel) and over smaller scales (mussel beds) (lower panel). The horizontal line shows the median, box shows the 25th and 75th percentiles, whiskers show the 5th and 95th percentiles, and circles are outliers. Filtration rate estimates are very approximate, and were calculated from population densities by assuming a filtration rate of 25 L/individual-day (Kryger and Riisgard 1988, Vanderploeg et al. 1995). These data should not be taken to represent typical densities of mussel populations, because biologists have usually chosen to study mussels at sites where mussels are abundant. From Økland (1963) Negus (1966), Magnin and Stanczykowska (1971), Tudorancea (1972), Haukioja and Hakala (1974), Lewandowski and Stanczykowska (1975), Fisher and Tevesz (1976), Green (1980), Strayer et al. (1981, 1994, 1996), James (1985, 1987), Miller et al. (1986, 1992), Hanson et al. (1988), Nalepa and Gauvin (1988), Holland-Bartels (1990), Huebner et al. (1990), Nalepa et al. (1991), the many studies compiled by Downing and Downing (1992), Ponyi (1992), Kesler and Bailey (1993), Miller and Payne (1993), Richardson and Smith (1994), Ziuganov et al. (1994), Balfour and Smock (1995), Gittings et al. (1998), Johnson and Brown (1998), Welker and Walz (1998), Soto and Mena (1999), Strayer (1999a), Hastie et al. (2000), Martel et al. (2001), Smith et al. (2001), Vaughn and Spooner (2004), and Lewandowski (2006).

(allochthonous inputs), and because of the existence of additional controls on food supply rates. These complications limit our ability to use simple models to predict how often and under what conditions unionoids are likely to control particle concentrations and composition. Nevertheless, it is possible to reach some limited conclusions. First, consider a well mixed stream with a concentration of edible suspended particles P as it first encounters a dense mussel bed. The water depth is z (technically z is the depth of the mixed zone, and could be less than the water depth if the water column is poorly mixed), the current velocity is v, and the density of mussels is D; following Kryger and Riisgard (1988), assume that each mussel clears the particles from 25 L/d. Further assume that the distance (and therefore time) under consideration is too short to allow for significant inputs of new edible particles from primary production or allochthonous sources. The particle concentration falls as the water passes over the mussel bed as follows

$$P(x) = P(0)e^{-2.9Dx10^{-7}/vz}$$

where x is distance (in m) from the upstream edge of the mussel bed. Fig. 43 shows solutions of this equation under a range of hydraulic conditions and mussel densities. Local depletion of food by a single mussel bed will occur only over dense mussel beds and in shallow, slow-flowing streams. For example, if the water is 10 cm deep and flows at 10 cm/sec, food will scarcely be depleted at mussel densities below $10/m^2$; and hydraulic conditions of 30 cm/s of current and 30 cm of depth will require mussel densities of at least $100/m^2$ to have much effect on food resources. Such high densities do occur (Fig. 42), but are not common. Of course, if the water column is poorly mixed, then food may be depleted over the mussel bed or even around individual mussels under less restrictive conditions.

Alternatively, consider a uniform, infinitely long reach of stream that is well mixed vertically and laterally and contains mussels along its course. In this case, particle inputs are not negligible, as in the previous case, but are key to determining the impacts of mussel feeding. As in the lake, we need to compare the rate of mussel grazing to the rate at which edible particles are generated. This situation can be modeled as a lake, with two key differences. First, streams typically receive appreciable lateral inputs of edible particles from allochthonous sources and by sloughing and resuspension from benthic primary production. Instead of comparing mussel

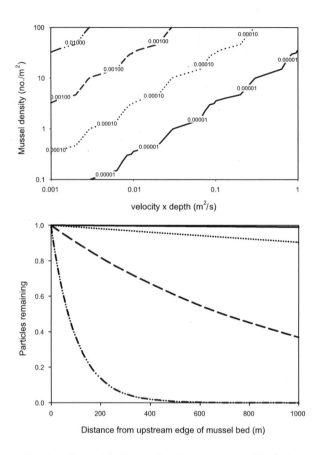

FIGURE 43. Results of a model of water flowing over a mussel bed. Upper. Aggregate clearance rate (proportion of the water column cleared per meter of stream) as a function of hydraulic conditions (i.e., water velocity times water depth) and mussel density. Lower. Loss of edible particles along the course of a mussel bed, for a range of aggregate clearance rates. The clearance rates in the lower panel are the same as those contoured in the upper panel.

clearance rates to phytoplankton growth rates, the comparison must now be with the sum of phytoplankton growth rates, allochthonous inputs, and sloughing from benthic sources. Unfortunately, these additional food sources make it difficult to use a simple analytical model to explore mussel impacts. Second, the positive feedback loop between the bivalve population and its food supply is less likely to occur in running waters than in standing waters because mussel feeding does not feed back onto the supply rate of allochthonous particles, and because algae in running waters often are limited by shading from silt or from an overhead canopy. This limitation will not be ameliorated significantly by unionoid grazing. Consequently, the per capita effects of unionoids may be more severe in running waters than in standing waters.

Despite these complications, it is possible to reach a few conclusions about conditions under which mussel grazing is likely to be large compared to the inputs of edible particles over long stream reaches (as opposed to individual mussel beds). Just as in lakes, the impacts of mussels on food resources will be greatest when mussel densities are high, when the water column is shallow, and when algal growth rates are low (i.e., in turbid, shaded, and nutrient-poor waters). The extent to which benthic production is suspended by currents or biological activity (e.g., Stenroth and Nystrom 2003), rather than being consumed by benthic grazers, will also influence whether mussels can deplete food resources. Finally, there may be interesting interactions with other stream-dwelling suspension-feeders (e.g., black flies, net-spinning caddisflies), which may themselves have high grazing rates (e.g., Wallace and Merritt 1980, Malmqvist et al. 2001).

These same considerations apply to benthic organic matter and deposit-feeding by unionoids, although the exact nature and supply rate of the food are even more difficult to define for these than for suspended food.

Third, unionoid individuals and populations may be food-limited if environmental conditions make it difficult for the animals to obtain food, whatever the ambient concentrations of food. This area has not been explored for unionoids, but has been addressed for other bivalves. For example, a high ratio of inorganic particles to organic particles has been shown to limit zebra mussel growth (Madon et al., 1998, Schneider et al., 1998). Likewise, strong or gusty currents may inhibit bivalve feeding (Wildish and Kristmansen 1997). We do not know if such environmental inhibitions often cause food-limitation in unionoids.

It is important to realize that food-limitation (of individuals) can occur if any of these three conditions is met at any time of the year, not simply for annual averages. Because food supply rates and environmental conditions vary widely through the year in both standing and running waters, transient food limitation is much more likely to occur than would be expected from an analysis of annual averages.

Thus, both empirical data and simple models suggest that unionoid populations may be food-limited, and point to the circumstances in which food-limitation is most likely. Nevertheless, we do not yet know if food limitation in nature is rare or widespread, light or severe. Several avenues of research might help us understand the nature and severity of food-limitation in unionoid populations. First, laboratory studies of the assimilability of various food items would help elucidate which parts of the heterogeneous mix of particles that unionoids capture are actually unionoid food. Likewise, laboratory studies of the long term performance of unionoids reared on different diets should continue to help us understand which parts of the unionoid diet are essential. If these laboratory studies are able to provide a clear definition of the quality of different kinds of food, it may be possible to take this information into the field and measure food availability in nature.

In the meantime, and in the event that laboratory studies are not able to provide a simple definition of food quality, it may be useful to use crude measures (e.g., chrorophyll *a* or particulate organic matter in a certain size range) as a working measure of food availability. These measures could be correlated with measures of individual performance (e.g., growth, fecundity) or population density as a first test of the strength of food limitation. I think it is especially important to look for evidence that anthropogenic eutrophication may have benefited unionoid populations in some circumstances. However, if food requirements turn out to be idiosyncratic, it may be difficult to identify food limitation in nature without detailed biochemical analyses of particles.

SEVEN

ENEMIES

Predation, parasitism, and disease are often thought to limit animal populations but these factors have not been thoroughly investigated for unionoids. We can list some of the predators and parasites that affect unionoids and we know something about the selectivity and feeding rates of a few predators. Nevertheless, we know very little about the geographic extent or ultimate demographic impacts of any enemy.

Mammalian predators (raccoons, otters, and especially muskrats) have received the most attention. These animals conveniently leave the empty shells of the animals they have eaten in neat piles along the shore, so it is possible to count how many mussels they've eaten, and compare the size and species composition of captured mussels with those of the mussel community from which they were taken. Many studies (e.g., Bovbjerg 1956, Hanson et al. 1989, Neves and Odom 1989, Convey et al. 1989, Watters 1994b, Jokela and Mutikainen 1995, Tyrrell and Hornbach 1998, Diggins and Stewart 2000, Zahner-Meike and Hanson 2001) have shown that muskrats may eat a lot of unionoids, and that they can be selective with respect to the size and species composition of the animals they eat (Table 11). Muskrat predation can take thousands of unionoids from local populations (Neves and Odom 1989, Hanson et al. 1989, Watters 1994b). Especially in small streams and ponds, they can almost eliminate mussel populations in a few years or less (Diggins and Stewart 2000, Zahner-Meike

TABLE 11 *Example of Species-selective Predation by Muskrats in the North Fork Holston River, Virginia*

Species	% in community	% in muskrat middens
Medionidus conradicus	36.2	18.7
Villosa nebulosa	20.1	21.0
Villosa vanuxemi	12.3	14.0
Pleurobema oviforme	10.2	15.2
Ptychobranchus subtentum	7.2	10.9
Lexingtonia dolabelloides	5.5	2.6
Fusconaia cor	5.1	12.8
Ptychobranchus fasciolaris	3.4	4.7

NOTE: From Neves and Odom 1989. The table excludes six species that were too rare to be collected in samples of the mussel community, but appeared in muskrat middens.

and Hanson 2001). The patterns of size-selectivity are not simple, with muskrats choosing larger-than-average animals from some populations and smaller-than-average animals from other populations. To the extent that muskrats feed on large, reproductive adults, they may have a disproportionate impact on the reproductive capacity of the mussel population. The patterns of species selection also are somewhat idiosyncratic, although the thin-shelled anodontines often are heavily preyed upon.

Optimal foraging theory probably could be used to make sense of patterns of muskrat predation. Zahner-Meike and Hanson (2001) reasonably suggested that muskrats may select mussels according to the benefit they receive (i.e., the amount of meat in the mussel) compared to the cost of capturing the mussel. They suggested that cost could be estimated by the mass of the mussel (meat plus shell); other terms that might be included in the cost term of a foraging model include the depth of the water from which the mussel is taken, the current speed, the distance from the mussel's location to the muskrat's retreat, and the difficulty of dislodging and opening the mussel. The last term presumably depends on the size of the mussel and its burrowing behavior; many anodontines burrow shallowly (and gape widely when removed from the water), for example, which may help explain why muskrats prefer them.

Nevertheless, it is impossible to estimate the demographic impact on muskrat predation on unionoid populations from the kinds of studies that have been done, for two reasons. First, as will be discussed shortly, the impact of predation on a population depends critically on the strength of density dependence in various demographic processes. Second, it is not clear from most of the studies that have been done whether mussels have a refuge from muskrat predation in the form of areas that are too deep or too far from shore for muskrats to forage. Alternatively, muskrats may stop seeking mussels once mussel densities become too low. For example, Jokela and Mutikainen (1995) showed that muskrat predation was intense in a band along the shoreline, beyond which the muskrats apparently did not forage (Fig. 44). Most predators will give up foraging as unprofitable before they literally eliminate the prey population, especially if foraging is specifically directed towards a particular prey item. The existence and size of such spatial or economic refuges will determine the extent to which muskrat predation threatens the viability of mussel populations. Refuges also influence whether muskrat predation is likely to shift the species composition of mussel communities, or whether muskrats eat all of the species, simply saving the least preferred species until last. Muskrats pose more of a threat to mussel populations if the body of water is small and if muskrats are foraging in the region for food items other than mussels than if the body of water is large or mussels are encountered only through special foraging trips.

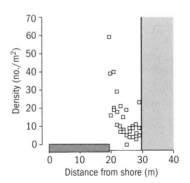

FIGURE 44. Example of spatial refuge from the effects of muskrat predation, from Jokela and Mutikainen (1995). Muskrats have eaten all of the mussels near the shore, but a large population remains in the center of the stream. Stippled area not sampled.

Other vertebrates (fish, birds, turtles) also eat unionoids (e.g., Adams 1892, Baker 1916, Coker et al. 1921, Fuller 1974, Berrow 1991), but their predation rates, selectivity, and ultimate impact on unionoid populations are not known. They usually are dismissed as quantitatively unimportant, although it seems possible that fish predation may be important in sites where molluskivorous species such as freshwater drum and large catfishes are abundant.

Several kinds of invertebrates eat mussels. Crayfish are important predators of snails (e.g., Perry et al. 1997, Brown 1998, Wilson et al. 2004b) and zebra mussels (e.g., MacIsaac 1994, Martin and Corkum 1994, Perry et al. 1997, 2000, Stewart et al. 1998, Reynolds and Donahoe 2001), and have recently been shown to eat small (<10-20 mm) unionoids (Klocker and Strayer 2004). The rapid spread and population outbreaks of invasive crayfish such as *Orconectes rusticus* in eastern North America, *Procambarus clarkii* in Africa, and *Orconectes limosus* in Europe may thus threaten unionoid populations around the world, although again nothing is known about crayfish feeding rates on unionoids in nature. Various microturbellarians are voracious predators of glochidia and small juvenile mussels in the laboratory and culture facilities (e.g., Coker et al. 1921, Delp 2002, Zimmerman et al. 2003). These animals are abundant (often $>10,000/m^2$) in lakes, streams, and rivers (Kolasa 2002), but nothing is known about their importance as predators of unionoids in nature. Finally, it seems likely that the many invertebrates that are generalized predators in freshwater sediments (e.g., chaetogastrine oliogochaetes (Coker et al. 1921), cyclopoid copepods, tanypodine chironomids and other insects) take juvenile unionoids, at least occasionally.

Parasites are widespread in unionoids and may have serious effects on their hosts. Digenetic trematodes castrate their unionoid hosts and may completely prevent mussel reproduction. Prevalence rates are usually >10% (sometimes >50%) in populations that have been studied (Jokela et al. 1993, Martell and Trdan 1994, Taskinen et al. 1994, 1997, Walker et al. 2001). It seems likely that the size of such heavily parasitized populations may be affected, although the demographic impacts of digenetic trematode infestations seem not to have been estimated. Aspidogastrid trematodes also are common in unionoids (Coker et al. 1921, Huehner 1984, Duobinis-Gray et al. 1991), with reported prevalence rates >10%. Their effects on individual mussels and mussel populations are not known. Ergasilid copepods parasitize mussel gills, but their effects are unknown. Prevelance rates can be very high (70–100%) in infested populations (Saarinen and Taskinen, 2004; Taskinen and Saarinen 2006). Unionicolid mites are another common parasite of unionoids, with prevalence rates sometimes >90% (Mitchell 1965, Edwards and Dimock 1988, Vidrine and Wilson 1991). Most of these mite species are highly host-specific, using only one or two species of mussels (Edwards and Vidrine 2006). Again, their effects on unionoids are not known. Finally, outbreaks of diseases (presumably bacterial and viral in origin) sometimes wipe out entire populations or communities of unionoids (Neves 1987), but even less is known about them than about the other predators and parasites mentioned here. Nearly all of these parasites or diseases have strongly differential effects on different species of unionoids, and so could potentially affect the composition of the mussel community.

Finally, humans often harvest unionoids for food, pearls, or mother-of-pearl (for buttons and ornaments) (Kunz 1898, Morrison 1942, Claassen 1994, Parmalee and Bogan 1998, Anthony and Downing 2001, Beasley 2001, Young et al. 2001). Whether humans collect the mussels by hand or use specialized tools such as brails, these fisheries are size- and species-selective, and often collect very large numbers of mussels. Whether because of gear limitations, economic incentives, or legal regulations intended to protect the stock, humans collect only large mussels. The lower size limit depends on the gear and the purpose of the fishery, but can be as small as 30 mm (e.g., Beasley 2001). In addition, the differential economic value of different species may affect the species composition of the harvest. Nevertheless, some harvest methods (e.g., brailing) may produce a large by-catch of species that are not the target of the fishery. Harvests can be immense (Claassen 1994, Anthony and Downing 2001), in some cases constituting essentially the entire assemblage of animals large enough to collect. Thus, early fisheries removed 100,000,000 mussels from a single mussel bed of 730,000 m^2 in the Mississippi River (which is equivalent to a *removal* of ~140 mussels/m^2) (Carlander 1954), and harvests from Illinois in 1913 alone (the peak of the pearl-button fishery) were 13 million kg of shells from live animals (Claassen 1994).

It does not require a formal analysis of these fisheries to conclude that human harvests of unionoids can strongly affect mussel abundance (Anthony and Downing 2001). Especially when modern gear is used or when the body of water is small or shallow, humans can remove nearly all of the mussels of reproductive age, at least for large-bodied species, over large areas. Because humans still would like to harvest mussels for mother-of-pearl or pearls, it would be desirable to be able to calculate a sustainable harvest for unionoid populations. The question of sustainable harvest has been addressed just a few times (Neves 1999, Anthony and Downing 2001, Hart and Grier 2004), but some states are now trying to manage mussel fisheries as sustainable resources (Neves 1999). Nevertheless, I have not seen a general formula for estimating sustainable harvest of unionoid populations; models from marine shellfisheries or finfisheries could be useful.

In addition to their direct effects as mussel predators, humans may affect the intensity of losses to other enemies. Fur trapping may reduce muskrat populations, at least locally; conversely, humans introduced muskrats to Europe, where they are now widespread predators of mussels. Humans have spread crayfish species widely outside of their native ranges (Hobbs et al. 1989, Lodge et al. 2000). Increasing nutrient inputs may have increased the densities of generalist predators. Finally, it is possible that activities such as ballast discharge and fish stocking have moved mussel diseases and parasites around the world, although this has not been studied.

Thus, we know that unionoids have some enemies (e.g., muskrats) that can kill or sterilize enough mussels to affect their distribution and abundance. We suspect that several other kinds of enemies (crayfish, flatworms, trematodes, and various diseases) probably have the potential to affect unionoid populations as well. Thus, it seems probable that any general theory to predict the distribution and abundance of unionoids must include the influence of enemies. Human activities such as muskrat trapping and introductions of muskrats and crayfish probably have caused the impacts of enemies to change over time in some places.

Nevertheless, we are far from understanding the severity or geographic extent of enemy impacts. Most studies of enemy impacts probably were done in areas where predation or parasitism was especially conspicuous, and so overestimate typical impact. Furthermore, we have only vague ideas about the times and places at which the impact of each enemy is likely to be the greatest. We know that predators and parasites often are species-selective, but do not know how often selective predation and parasitism changes the composition of unionoid communities.

Most seriously, studies of unionoid enemies rarely have been put into a demographic context, so we have very little idea of how observed impacts of enemies translate into changes in population size, geographic range, or demographic characteristics of unionoid populations and communities. If a mussel population is near the threshold of viability, even a little added predation may drive it to extinction. Conversely, if key controls exhibit strong negative density-dependence, then even substantial predation may have little effect on the population. Predation itself probably is typically density-dependent, with strong density-dependence in specialized predators and weaker density-dependence in generalized predators. Likewise, other controlling processes (e.g., immunization of host fish, quality or quantity of food) may be density-dependent. As a result, statements such as "predator X consumes Y unionoids" provide little insight into the demographic effects of the enemy. An accurate assessment of the effects of enemies on the size or viability of unionoid populations will require putting losses from enemies into a life-table analysis or some other sort of demographic model.

EIGHT

IMPLICATIONS FOR CONSERVATION

It is worth emphasizing that human activities have had major effects on all five of the classes of factors that control unionoid populations, and thus have had large, varied impacts on unionoid distribution and abundance (Fig. 45). Model and empirical analyses (Figs. 11, 12, 15) suggest that the barriers that humans have spread throughout many river systems may have very large impacts on mussel metapopulations, and that the full long-term effects of these barriers probably have not yet been realized. There are now more than 45,000 large dams, which probably pose absolute barriers to mussel dispersal, as well as perhaps one million smaller dams, on the world's streams and rivers today (Jackson et al. 2001, Malmqvist and Rundle 2002). I have not seen any estimates of the numbers of other barriers (e.g., long stretches of habitat unsuitable for fish or mussels), but barriers other than dams are numerous in some river systems. In addition to their impacts on the functioning of mussel metapopulations, these barriers will hinder any attempts by mussel species to adjust their ranges to changing climate.

Human activities have changed the physicochemical habitat in lakes and especially in running waters in countless ways. We have changed the hydrology (e.g., Nilsson et al. 2005), sediment (e.g., Waters 1995) and nutrient (e.g., Carpenter et al. 1998) loads, thermal regimes, and light regimes (e.g., through the removal of riparian vegetation), physically modified shorelines, dredged channels, filled shallow waters, and polluted waters

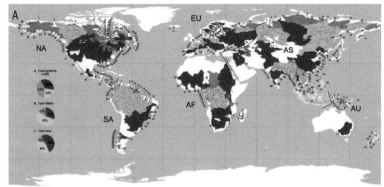

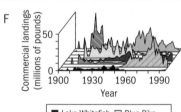

with toxins. Essentially all of the fresh waters in developed regions have been affected by these changes to some degree, and lakes and streams even in the most remote regions may be affected through pathways like atmospheric deposition. Again, because of the slow response times of some parts of the ecosystem and the long life-spans of mussels, the full effects of these large, pervasive changes to habitat will not be expressed for many decades.

The same human-made changes to dispersal and habitat quality that have affected mussels directly have also had very large effects on populations of their fish hosts (e.g., Trautman 1981). We also have introduced many freshwater fish species outside their native ranges (e.g., Fuller et al. 1999, Rahel 2000), carried fish diseases around the world (Bartholomew and Wilson 2002), and heavily harvested many stocks of freshwater fish, some to the point of extirpation (e.g., Nepszy 1999, Allan et al. 2005). Consequently, very few bodies of water in Europe, North America, and Australia retain their original fish communities. Because it seems likely that at least some mussel populations are controlled by host availability, these large human-caused changes to freshwater fish communities probably have had strong effects on mussel communities which will not be fully realized for decades.

Large, widespread increases in loading of phosphorus and nitrogen to fresh waters and transformations of watersheds have led to large changes in the amount and quality of phytoplankton (Carpenter et al. 1998) and other food available to mussel populations around the world. Organic

FIGURE 45. Opposite. Examples of the large, pervasive anthropogenic changes to freshwater ecosystems that probably have affected mussel populations. A. The extent of damming on the world's large river systems. Basins shown in dark gray are strongly impacted (by channel fragmentation and flow regulation), basins shown in medium gray are moderately impacted, and basins shown in light gray are unimpacted (Nilsson et al. 2005). B. A large dam in Tennessee; there are ~45,000 such large dams around the world today; C. An example of severe habitat change to a stream (Hog Creek, Ohio); D. An example of severe change to the watershed, in this case the watershed of Big Darby Creek, Ohio, a hotspot of remaining unionoid diversity; E. The black carp (*Mylopharyngodon piceus*), a molluskivorous fish from Asia that may become established in the wild in North America as a result of aquacultural operations (Leo G. Nico, U.S. Geological Survey); F. Changes to the fish community of Lake Erie, as reflected by catches of the commercial fishery (Baldwin et al. 2007).

wastes from human activities also probably contribute to the unionoid diet in many places. The implications of these massive changes to the unionoid food base have not been explored, but could be substantial. Invasive species such as the zebra mussel and *Corbicula* have been widely introduced outside of their native ranges, and have greatly depleted food resources for unionoids in some lakes, rivers, and streams. Unionoid populations surely have been damaged by these depletions, as well as by mechanical interference from the introduced bivalves.

As well as serving as a direct enemy of unionoids through harvesting for mother-of-pearl, pearls, and food, humans have altered populations of other enemies of unionoids. Muskrats have been introduced into Europe, and rusty crayfish (*Orconectes rusticus*) and other predatory species have been widely introduced throughout North America, Europe, and Africa. The molluskivorous black carp (*Mylopharyngodon piceus*) is on the brink of establishment in North America, where it could become common and widespread (Nico et al. 2005). It also seems likely that humans have spread diseases or parasites of unionoids outside their native ranges, although this has not been demonstrated.

In sum, the contexts in which mussel populations operate have been completely changed by human actions. Attempts to explain mussel distribution and abundance must include humans, and attempts to understand or manage the effects of humans on mussel species must consider all five classes of controlling factors. Furthermore, these changes in unionoid populations presumably have consequences for ecosystem functioning and for the distribution and abundance of other suspension-feeders in freshwater ecosystems, but these cascading consequences have scarcely been considered. Finally, it seems likely that contemporary mussel populations are far from equilibrium with respect to past human impacts, so that the full effects of past human actions will not be seen for decades to centuries.

WHAT FACTORS CONTROL DISTRIBUTION AND ABUNDANCE OF UNIONOIDS?

I hope that I have convinced you that each of the five factors that I have discussed—dispersal, habitat, fish hosts, food, and enemies—actually limits the distribution and abundance of unionoids at some times and places. It seems impossible to neglect any factor as irrelevant or unimportant.

Further, I think it is clear that these five classes of controlling factors at least have the potential to interact strongly with one another. In particular, note that there is some substitutability across controlling factors. For instance, low predation rates could compensate for high juvenile mortality from poor habitat conditions or low fecundity of food-limited adults, or excellent survival of juveniles might make up for a scarcity of fish hosts (Geist et al. 2006). I also conclude that although we know a great deal about unionoid ecology, we do not have all the critical information we need about any of the five factors. In particular, the following seem to me to be especially critical needs:

- Estimates of past and present dispersal rates within and across drainage basins are needed to assess whether human-made barriers in stream systems are likely to lead to local and global loss of unionoid species, to estimate the speed of such species losses, and to determine whether unionoids will be able to adjust their ranges in response to climate change.
- More extensive tests of the importance of sediment stability in limiting unionoid populations, and development of practical, robust methods to estimate sediment stability would be highly desirable (cf. Lamouroux et al. 1992, Morales et al. 2006).
- Field surveys of interstitial unionized ammonia are needed to determine if excessive ammonia has caused recent declines in unionoid populations.
- We need more bioassays to determine the extent to which residues from past episodes of pollution have resulted in sediments that are toxic to juvenile unionoids, and thereby prevented the recolonization of unionoids to sites that are otherwise suitable.
- It would be useful to apply demographic tools to estimate the effect of mortality from toxins or predators on the size and viability of mussel populations.
- It would be helpful to measure the strength of fish immune responses and the extent of immunity in natural fish populations, to determine whether density-dependent depletion of hosts regulates the size of mussel populations or causes interspecific competition.

- We need critical studies to determine if, how often, and under what circumstances mussel populations are limited by the availability of hosts.
- More information on the effects of food quantity and quality on growth and fecundity of unionoids would be desirable, ultimately to be linked to the demographic response of the mussel population. Information on the effects of widespread human-caused eutrophication would be especially important (eutrophication may also increase densities of fish hosts).
- More extensive, unbiased information on the prevalence and effects of parasites is needed to assess whether control by enemies is likely to occur, and whether it is strongly density-dependent.

Despite the length of this list and the difficulty of collecting some of this information, I believe that it probably would be feasible to collect most of this missing information if we are sufficiently motivated to predict mussel distribution and abundance. Thus, I conclude that a satisfactory theory of unionoid distribution and abundance will have to consider all five classes of factors and at least the possibility of interactions. We will need to call in Dr. Frankenstein.

PART 3

MAKING THE MONSTER WALK

NINE

THREE MODELS FOR MUSSEL ECOLOGY

It seems clear that unionoid distribution and abundance is controlled by multiple factors, and that we therefore need some sort of plan for building a working theory that includes the effects of these multiple factors. It is not just unionoid ecology that needs a Dr. Frankenstein; many ecological variables are simultaneously controlled by multiple factors. Furthermore, the integration of multiple controls into a working theory is not a trivial problem. Consider the acrimonious, decades-long controversy about bottom-up (i.e., nutrient) "versus" top-down (i.e., food-web) control of phytoplankton biomass and production in lakes. Historically, phytoplankton biomass was thought to be controlled largely by nutrients (most often phosphorus; e.g., Likens, 1972, Dillon and Rigler 1974). In the 1980s, the idea that food web structure could control phytoplankton biomass (e.g., piscivorous fish suppress planktivorous fish, thereby releasing algivorous zooplankton and suppressing phytoplankton) became popular (e.g., Carpenter et al. 1985; Carpenter and Kitchell 1996). A great deal was written about whether nutrients *or* fish controlled phytoplankton biomass, with relatively little discussion of the various ways in which these two classes of controls might jointly control phytoplankton (e.g., DeMelo et al. 1992, Carpenter et al. 1995, Pace et al. 1999, Wetzel 2001, pp. 464–466). Lake pelagic zones are one of the simplest, best-studied, and best-understood of all the ecosystems on Earth, yet ecologists were not able to easily agree how these two classes of factors controlled a single, relatively simple variable. If the problem of assessing and integrating just two classes of variables has occupied aquatic ecologists for decades, think how much more difficult these tasks will be in more complex, less well-studied ecological systems. Thus, I believe that the problem of integration of multiple controlling factors is a general and difficult problem that many ecologists will need to confront, sooner or later.

How can we integrate multiple controlling factors into a satisfactory understanding of mussel populations? We need to develop some sort of quantitative model to bring together the different controlling factors. At the risk of oversimplification, there seem to me to be two broad classes of integrative models. The first is mechanistic, in which we seek to develop an integrative theory whose pieces are themselves small working theories about how each process works. That is, we use the detailed information that we have gathered about how each of the multiple controlling factors (dispersal, habitat, hosts, food, and enemies) works to provide the functional relationships and parameters of our final quantitative model that predicts unionoid distribution and abundance. This seems to me to be closest to the classical reductionist view of science, and it is my impression that most working ecologists see their work as contributing to the development of such a mechanistic model. For example, in discussing the problem of coastal eutrophication, Cloern (2001) stated "The ultimate objective of all this hard work is a mechanistic understanding, based on scientific principles, from which management strategies can be designed. . . ." This approach demands that Dr. Frankenstein build an actual living human from body parts.

Alternatively, we may use a more empirical approach, in which the detailed information that we have gathered about individual processes is used to inspire or guide the construction of an integrative model, but does not literally serve as the pieces of the overall model. In this approach, it as if Dr. Frankenstein is trying to build not a real human, but rather a mannequin that serves some of the functions of a living human. He may be inspired to give the mannequin an artificial hand whose structure and function are based on a human hand, but he will use whatever materials are available rather than a real human hand. The mannequin may be very simple if the function is very simple (providing a place to hang a shirt), or it may have to be very complex if we are trying to do a complex task (artificial conversation with convincing facial expressions).

It seems to me that ecologists have either implicitly or explicitly adopted three general approaches, two mechanistic and one empirical, in dealing with the problem of multiple controlling factors. I will briefly evaluate each of these approaches.

A SIMPLE MONSTER

One common approach has been to assume that a single controlling factor logically must have precedence, so that other potential controls can be ignored. This approach, Liebig's Law of the Minimum, exemplified by Liebig's wonderfully evocative image of a leaky barrel, is simple, tractable, and useful in some circumstances. It is literally valid only when the multiple controlling factors do not interact with one another and are homogeneous across space and time. In this case, it would be possible to calculate the expected abundance of mussels independently based on each of the five controlling factors: the predicted abundance of mussels N is simply the minimum of these estimates. Thus,

$$N_{dispersal} = f(dispersal)$$
$$N_{food} = f(food)$$
$$N_{enemies} = f(enemies)$$
$$N_{habitat} = f(habitat)$$
$$N_{hosts} = f(hosts)$$
$$N = \min(N_{dispersal}, N_{food}, N_{enemies}, N_{habitat}, N_{hosts})$$

where each of these equations is based on detailed studies of each process.

This approach to integration would seem to be of limited utility for unionoids. Most obviously, its assumptions of non-interacting, spatially and temporally homogeneous controlling factors are never literally met in nature. Controlling factors interact. For instance, high predation rates may be counteracted by high fecundity resulting from plentiful food, or by habitats with refuges from predation, or by high rates of successful glochidial transformation from an abundant host pool. Mussel habitats aren't homogeneous. It is difficult to think of an environment where the assumptions of spatial and temporal homogeneity are violated more egregiously than in a typical stream (Fig. 46), and even lake bottoms are conspicuously

FIGURE 46. Two photographs of Webatuck Creek, New York, showing the characteristic spatial and temporal heterogeneity of mussel habitats. The upper picture shows the creek during low water in September 2007, and the lower picture shows the same reach of creek in September 1999.

heterogeneous. Thus, the minimum N_i calculated from a point estimate (whether a mean, a median, a maximum, a minimum, etc.) of conditions in a body of fresh water would not be expected to give a satisfactory estimate of actual mussel population densities (Strayer et al. 2003) (although an empirically derived estimate may suffice; e.g., Lamouroux et al. 1992).

A more technical problem with this approach is that mussel ecologists haven't thought much about predicting mussel densities from dispersal and habitat, but rather have treated these as producing binary responses: either dispersal and habitat allow a mussel population to be present or constrain it to be absent. That is, $N_{dispersal}$ and $N_{habitat}$ are either 0 or 1. This limitation could be accommodated in the Liebigian world; if either $N_{dispersal}$ or $N_{habitat}$ is 0, then mussels are absent; if they are both 1, then N might be estimated as min $(N_{food}, N_{hosts}, N_{enemies})$.

Despite these shortcomings, the Liebigian approach may be useful in a heterogeneous, interactive world in at least two cases. First, it may suffice to predict the presence or absence of mussels. Mussels probably will be absent from sites where one (or more) of the N_i are estimated to be zero or highly negative. For instance, suppose that a mussel species needs a phytoplankton biomass of 5 μg chl a/L to survive, but that algal concentrations in a stream rarely exceed 3 μg chl a/L. It is likely that mussels will be absent from this stream, regardless of the quality of the habitat, the abundance of fish hosts, the absence of predators, and the existence of dispersal corridors. Likewise, we could predict mussels to be present at a site where all N_i are highly positive. Nevertheless, strong interactions among controlling factors, or strong spatial or temporal variation in controlling factors may invalidate even these simple conclusions, so this simple approach is most likely to be useful only as a first-order screening procedure.

Second, Liebig's approach may be useful when one of the controlling factors is strongly constraining and the others are benign (i.e., when one of the N_i is much smaller than all of the others), interactions are small (i.e., there is limited substitutability among controlling factors), and heterogeneity is not pronounced. In such a case, the Liebigian approach should give a reliable estimate of population density. These conditions are most likely to be met when the approach is applied over a limited geographical and ecological range. For instance, one might examine the effect of varying host fish abundance in a series of streams with similar habitat conditions and food resources in a small geographical area. As the domain of the model is expanded to include a wider range of ecological conditions, the model is increasingly likely to fail.

To see what would be needed to produce a more realistic model, consider in turn the three chief ways in which the Liebigian approach fails as a literal description of the real world: the real world is spatially and temporally heterogeneous, and controlling factors may interact.

SPATIAL HETEROGENEITY

The world that mussels inhabit is spatially heterogeneous with respect to all of the potential controlling factors (e.g., Figs. 17, 19, 20, 22, 44, 46), violating an assumption of the simple Liebigian approach. Even when a single controlling factor is considered, spatial heterogeneity can invalidate simple models if nonlinearities in governing equations or interactions among patches are important (e.g., Strayer et al. 2003). Furthermore, the controlling factors usually differ widely in the pattern and scale of their spatial variation. Thus, water chemistry may be relatively homogeneous over km-long reaches of stream, whereas host fish abundance may vary greatly over scales of 1–100 m, and interstitial food or dissolved oxygen may vary strongly on a scale of cm. This spatial variation introduces serious complications to models of mussel distribution and abundance.

These complications are perhaps best appreciated from a simple example. Consider the situation shown in Fig. 47, in which a small, stable gravel bar is surrounded by shifting sands. Mussels are abundant on the gravel bar but are absent from the sand. It is clear enough that a model based on average sediment grain size or average stability of the reach is unlikely to be satisfactory. A more serious problem is that the information shown in Fig. 47 is consistent with several mechanisms of population control. The most obvious interpretation of Fig. 47 is that the mussel population is habitat-limited, and that certainly is a possibility. That is, we could increase the size of the mussel population if and only if we increased the amount or quality of habitat. An alternative interpretation is that the mussel population is jointly controlled by habitat and some other factor, say, host availability. Perhaps habitat controls the spatial extent of the mussel population, while host availability controls the density of mussels per area of habitat. If this interpretation is correct, we could increase the size of the mussel population by either improving habitat or increasing the number of hosts. Finally, and perhaps most disturbingly, Fig. 47 is consistent with the interpretation that the size of the mussel population is controlled solely by a factor other than habitat, say, host availability. Perhaps hosts are so scarce that only 20 mussels can be supported in the whole study area. Habitat determines where those mussels live, but not how many mussels there

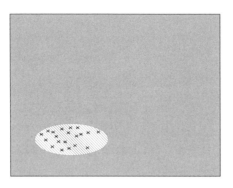

FIGURE 47. Schematic diagram of a bed of mussels (black x's) living on a stable gravel bar (cross-hatching) surrounded by shifting sand (solid gray).

are. If this interpretation is correct, then improving habitat will not increase the size of the mussel population—it will merely change its spatial distribution. Only increasing host availability can increase the number of mussels. There is no way to decide which of these interpretations is correct merely by examining the evidence in Fig. 47. A real example of exactly this problem is the debate over whether artificial reefs increase fish populations or merely change the spatial structure of fish populations by attracting fish from other habitats (e.g., Bohnsack 1989, Grossman et al. 1997).

Another common complication is that movement of organisms and materials allow some "spillover" of conditions in one patch into nearby patches. For example, dispersal of organisms across a spatially heterogeneous environment may allow animals to live in habitats that are not actually capable of supporting a viable population over the long term ("sink" habitats) (e.g., Grossman et al. 1995).

In addition, if the world is spatially heterogeneous, we need to be careful to specify the scale at which we are trying to predict mussel populations. Models to predict density of mussels at the 1 m² level within a 1000 m² sand bar in a lake will generally have very different structure, parameters, and predictive success than models to predict the mean lakewide density of mussels in a series of lakes across North America, for example.

The pronounced spatial heterogeneity in controlling factors (and in mussel populations themselves) raises the following questions: (1) under what circumstances do we need to model this spatial variation explicitly to effectively predict mussel populations? (2) what sorts of descriptions and models of spatial heterogeneity will be most effective? (3) at what spatial scale (grain and extent: Turner et al. 2001) is spatial heterogeneity best dealt with? These difficult questions are currently at the cutting edge of ecological research (e.g., Turner et al. 2001; Lovett et al. 2005) and do not appear to have simple general answers.

If we knew the spatial extents over which the mussel population and its controlling factors operated, we might be able to model the functioning of the mussel population by using a piecewise approach to modeling processes within and between patches at the largest spatial extent relevant to the resources or mussel population. That is, if fish were an important limiting factor, we could divide the study area into a series of patches corresponding to the home range of fishes, then calculate the $\min(N_i)$ for each factor within each patch and estimate N as the sum of the $\min(N_i)$ across all patches. However, we still know little about the spatial extent of interactions in mussel communities, and the general subject of defining spatial domains of ecological interactions is also a difficult problem that is just beginning to be explored (e.g., Caley et al. 1996, Power and Rainey 2000, Finlay et al. 2002, Polis et al. 2004, Power 2006).

TEMPORAL HETEROGENEITY

Temporal heterogeneity may pose an even greater problem than spatial heterogeneity for modeling unionoid populations. Essentially all of the issues raised by spatial heterogeneity apply to temporal heterogeneity as well. Because of the long life span of unionoids and the slow response time of many aspects of freshwater ecosystems (e.g., sediment routing), the influences of one event on subsequent events may last for many decades. Furthermore, the temporal shadows cast by different kinds of events will have different lengths: remember, for example, that the effects of dam construction on mussel populations in the Upper Mississippi and Cumberland Rivers in the early to mid-20th century have still not yet been fully expressed (Heinricher and Layzer 1999, Kelner and Sietman 2000), whereas the effects of a transient algal bloom on mussel growth probably are undetectable after a few days. Thus, as was the case for spatial heterogeneity, the temporal patterning of different controlling factors and the persistence of their effects will be non-concordant. Likewise, we will need to decide when it is desirable to include temporal heterogeneity explicitly in our models, and when it can be ignored or simplified.

One change that could help greatly in dealing with temporal heterogeneity and lags would be collection of data that could serve as leading indicators of the status of mussel populations, instead of data only on the presence or population density of adult mussels. The difficulty with collecting information only on adult mussels is that the density of adult mussels depends strongly on past ecological conditions as well as current conditions. For instance, it is impossible to distinguish populations that were viable 50 years ago and are now dying out from populations that are currently viable simply by comparing the densities of adults. It would be very helpful to have indicators of population status that could be used to diagnose the current demographic status of the population, and allow for better projections of future viability. The most obvious candidates are the traditional variables of population biology: survivorship and fecundity schedules. The advantages of such variables are that they are the subject of large theoretical and empirical literatures, and are used widely in models of population dynamics and viability. Such data are not always simple to collect, however, and few examples have been published for unionoids (e.g., Bauer 1983, 1987a, Young and Williams 1984a, Jansen and Hanson 1991, Hochwald 2001, Hart et al. 2001, Haag and Staton 2003, Villella et al. 2004). Some authors (e.g., Monroe and Newton 2001, Newton et al. 2001) have noted that physiological variables (e.g., energy stores) may also provide useful information on the current condition of mussels, although we need research on how best to choose and interpret such physiological variables.

INTERACTIONS AMONG CONTROLLING FACTORS

There are many strong interactions among the five classes of controlling factors. To see this, consider just the ten possible two-way interactions among these five classes, as enumerated and briefly discussed below. Dispersal is often thought of as being relatively non-interactive with other controlling factors, and therefore best treated at a higher hierarchical level (e.g., Guisan and Thuiller 2005). Nevertheless, there are examples of important interactions between dispersal and other factors that control unionoid populations. For example, (1) the extent and spatial distribution of suitable habitat surely affects dispersal between parts of a unionoid metapopulation, which may have strong consequences for patch occupancy and metapopulation persistence (see above). Dispersal interacts with host availability (2) because the abundance and species composition of the host pool must affect dispersal rates. Likewise, the effectiveness of barriers to dispersal, whether natural or human-made, will depend on the abilities of the hosts to disperse and surmount barriers. Although local dispersal of adult mussels may depend on food availability (Bovbjerg 1956), it seems unlikely that interactions between food and dispersal (3) are generally important. Barriers to dispersal may accentuate the effects of enemies (4) by preventing the re-establishment of local populations eliminated by enemies, or the augmentation of marginal or sink populations injured by enemies. This interaction may contribute to the long-term damage of the many barriers that humans have placed in river systems.

There are potentially important two-way interactions among all of the factors other than dispersal as well. Habitat may interact with host availability (5); changes in habitat may affect the distribution or behavior of hosts, thereby changing their overlap with mussel glochidia or the location into which juvenile mussels are dropped. Changes in host population size may affect the local distribution of hosts as well, with similar effects on glochidial contact and the sites of juvenile settlement. We have already seen (Fig. 21) that current speed and food concentration interact (6) to determine the rate of food delivery to suspension-feeders such as adult unionoids. There may be important, although unstudied, interactive effects on juvenile unionoid survival between benthic organic matter and habitat. High benthic organic matter may be beneficial to juveniles when sediments are highly permeable, but cause anoxia and juvenile mortality when sediments are not so permeable. Habitat structure will determine the extent to which there are refuges against enemies (7) (e.g., water too deep or far from shore for muskrats to forage, water too swift for crayfish), and thereby modulate their effects. Host availability and food may be substitutable to some extent (8): high production of glochidia from well-fed mussels may be able to compensate for low glochidia-host contact rates, for example. Furthermore, high environmental productivity may increase both host availability and mussel fecundity, leading to a doubly positive effect of productivity on mussels.

Finally, there it at least the possibility of trophic cascades from host fish (Carpenter and Kitchell 1996), leading to partial control of a mussel's food by its hosts. I doubt that interactions between hosts and enemies (9) are usually strong, but there are a few plausible examples. It has been known for a long time (e.g., Coker et al. 1921) that the freshwater drum (*Aplodinotus grunniens*), a specialized molluskivore, serves as a host for many mussel species (it is a known or suspected host for 15 species (Cummings and Watters 2005)) and is often heavily infested with glochidia. It is thus both a host and an enemy of unionoids; other molluskivorous fish may play a similar dual role. Host fish may suppress populations of enemies; for example, large centrarchids such as black bass (*Micropterus* spp.) can control crayfish populations (e.g., Rabeni 1992, Englund 1999, Maezono and Miyashita 2004). Host fish may be intermediate or definitive hosts of the digenetic trematodes that castrate unionoids (see discussion above), so the dynamics of a mussel's enemy population may be linked to that of its host population. Note also that humans play a dual role as an enemy of mussels and as controllers of host populations. The demographic effect of an enemy will depend on productivity or replacement rate of the mussel population, and therefore its food resources (10). There may be apparent competition effects (Holt and Lawton 1994) when an enemy preys on a competitor, as is the case when muskrats eat *Corbicula* (Neves and Odom 1989) or crayfish eat zebra mussels (Perry et al. 1997, 2000). Conversely, there may be spillover effects from a shared predator when high productivity increases the availability of alternative prey and thereby increases the population of a unionoid predator.

I hope that this tedious recitation has convinced you that interactions among controlling factors are not merely a theoretical possibility, but are widespread, varied, and probably often strong. Higher-order interactions among the controlling factors surely exist as well, and may be important, but I will not discuss them here because we know so little about them.

Interactions among controlling factors can have different functional forms. Thus, it will not always be easy to decide on the best way to model interactions, even when we recognize their existence. Finally, I note that the existence of important interactions and the difficulty of modeling them are by no means restricted to unionoid ecology, but are very general features of ecological systems, even though they have not been much studied (but see Cloern 2001, Pomeroy and Wiebe 2001, Moore et al. 2004).

A REAL MONSTER

It is perhaps natural to respond to the problems of a simplistic Liebigian approach by trying to build a comprehensive model that includes all of the effects of each of the five classes of controlling factors (including their interactions) on unionoid populations. I believe that this is neither feasible nor necessary. A really comprehensive, mechanistic model of unionoid populations would have to include information about all of the processes that affect unionoid populations—dispersal, habitat, host availability, food, and enemies—as well as their interactions. This information would have to be collected across heterogeneous environments, and over long periods of time. It would be a daunting task to collect this information for even one population, let alone the many populations needed to parameterize and test a model.

Furthermore, it would almost certainly be even harder to construct and test a model than to collect the data. There are a large number of relationships connecting the five classes of controlling factors with a focal unionoid population. Thus, in its most general form,

Number of mussels = f(dispersal, habitat, hosts, food, predators, dispersal*habitat, dispersal*hosts, dispersal*food, dispersal*predators, habitat*hosts, habitat*food, habitat*predators, hosts*food, hosts*predators, food*predators, dispersal*habitat*hosts, dispersal*habitat*food, dispersal*habitat*predators, dispersal*hosts*food, dispersal*hosts*predators, habitat*hosts*food, habitat*hosts*predators, habitat*food*predators, dispersal*habitat*hosts*food, dispersal*habitat*hosts*predators, dispersal*hosts*food*predators, dispersal*habitat*food*predators, habitat*hosts*food*predators, dispersal*habitat*hosts*food*predators).

Surely some of these terms could be neglected, but there is no way a priori to know which terms may be safely discarded. Each of these relationships could be represented by a wide range of mathematical functions and parameter values. It is impossible to know which of these model structures and parameters is most nearly "correct" without extensive testing of the model parts and the entire model. These are familiar problems with large mechanistic models in ecology, and they have been discussed in detail by others (e.g., Linhart and Zucchini 1986, Hakanson 1995, Hilborn and Mangel 1997, Oreskes 2003). I therefore conclude that it is not feasible to construct a really comprehensive, mechanistic model to predict the distribution and abundance of unionoid mussels.

Nevertheless, it is this sort of comprehensive model that is implicitly used to justify the unfocused collection of data on any topic that might conceivably affect the performance of mussels or mussel populations. Rigler and Peters (1995) suggested that unfocused collection of any data that might remotely be considered to be useful is a major problem in contemporary ecology. This unfocused effort is inefficient at best, wasting our too-scarce resources, and can be ultimately futile if the integrative model cannot be built.

HOW MUCH PREDICTIVE POWER DO WE NEED?

Another important reason for not trying to build a comprehensive mechanistic model is that we neither require complete predictive power nor do we have unlimited resources. Thus, our goal in model building is more likely to be to achieve some desired level of predictive power at the lowest cost, or to achieve the highest predictive power possible given the resources that are available to us, rather than to explain every detail about unionoid distribution and abundance. Either criterion should lead us to emphasize efficiency over completeness in model building. In such a situation, I suggest that it may be best to start with simple models, then systematically add complexity as needed. The order by which we add various pieces to the model may be determined by the amount of predictive power that we think they will add, or by the cost or availability of the information we need. As complexity is added, we need to evaluate carefully whether the added complexity meets our needs for predictive power, and whether it is a practical and economical way to obtain predictive power.

If we embrace this approach, then we may evaluate and adopt models that are patently unrealistic. The key criteria for using a model are its predictive power, cost, and ease of use, not its degree of realism or completeness.

EMPIRICAL APPROACHES

Thus, a third common approach to dealing with multiple controlling factors in ecology is empirical modeling. Empirical models use whatever data is available and thought to be a useful to predict the variable of interest, without worrying about whether explicit or parameterized mechanistic links can be drawn between the independent and dependent variables. Empirical approaches have been used very widely in ecology (e.g., Peters 1986), and have some obvious advantages over more mechanistic approaches. It is often relatively easy to build an empirical model for a specific situation using whatever information is available. Missing processes and mechanisms can be accommodated through the use of surrogate variables and fitted parameters. There are of course important limitations on the construction and interpretation of empirical models, as we will see later.

To appreciate the capabilities of empirical models, it may be useful to consider three hypothetical examples of very different applications of empirical models to predict mussel population densities (Fig. 48). The most straightforward empirical models are probably those in which we have data on some of the factors that we think directly affect mussel populations. For instance, we might be trying to model the abundance of a mussel species at sites across a 3000 km² watershed from data on physical habitat variables (slope, width, roughness); the list of fish species that have been

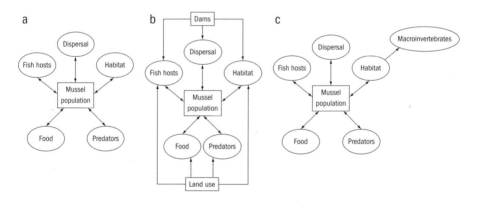

FIGURE 48. Examples of different structures of empirical models for predicting unionoid mussel population density. In (a), the model is built using information on controlling factors for which data are available; other controlling factors (dispersal and food) are not explicitly included; in (b) the model is built using variables that are thought to be related to the proximate controlling factors, rather than the proximate controlling factors themselves; in (c), the model is built using variables that themselves driven by proximate controlling factors, and are therefore thought to covary with mussel populations.

collected at each site; and the presence or absence of muskrats (Fig. 48a). Unlike mechanistic models, though, we are not trying to capture the actual functional relationships between these variables and mussel populations (i.e., the fitted coefficients may not correspond to any identifiable processes such as the daily consumption rate of muskrats).

Alternatively, we might be interested in the effects of human actions that affect the processes that control mussel populations; i.e., controlling factors that are a step removed from the mussel population (Fig. 48b). For instance, there is a lot of interest in the effects of land use and impoundments on mussel populations. Thus, we might try to model whether successful recruitment of a particular mussel species is taking place in a large number of sites across a 25,000 km^2 ecoregion as a function of land use and upstream impoundments. The independent variables in this case might be land use in the watersheds and along the riparian corridors of each study site, as well as the distance to the nearest upstream dam and its live storage capacity. The multifarious relationships that connect land use or impoundment to mussel populations are implicit in this model, but do not concern us directly and do not need to be specified.

As a third example, we might try to predict mussel presence or abundance using surrogate variables (i.e., variables that do not affect mussel populations directly, but which are correlated with controlling variables). For instance, suppose we are trying to determine if removing a dam will increase the range of a mussel species; i.e., are there large areas above the dam capable of supporting mussel populations? We might have data on macroinvertebrate community composition at many sites above and below the dam. The macroinvertebrate community probably doesn't have important direct effects on the mussel population, but it may have close relationships with the habitat variables, host community, and food that do matter to the mussel (Fig. 48c), so it is reasonable to think that we might be able to model mussel population extent or density from these variables. A real situation similar to this example was reported by Baldigo et al. (2002), who used a wide range of data, including macroinvertebrate community structure, to assess whether dam removal was likely to increase the local distribution of an endangered mussel.

As these three examples show, empirical models can be highly varied to match specific research or management questions, can be tailored to fit the data that are actually available, and are relatively unconcerned with literally replicating pathways of interaction. Many statistical tools (e.g., multiple regression, logistic regression, and many others) can be applied in empirical modeling, depending on the nature of the data, the research or management question, and the shapes of statistical relationships among variables.

Nevertheless, empirical models must be constructed and interpreted carefully. Probably the most frequent objection to empirical models is that they are based merely on correlation rather than on an understanding of the causative links between variables. One implication of this criticism is that knowledge of causative mechanisms is somehow intellectually superior to other kinds of knowledge. If mechanistic understanding is important to us, then purely empirical approaches may have serious flaws. If our goal is predictive power, however, this criticism need not concern us.

More seriously, models based on correlations may not give reliable results when they are not tested properly or are applied outside the domain of conditions for which they were developed. Thus, a model of mussel abundance developed for lakes in New York may not work well when applied to lakes in Michigan. Likewise, a model of unionoid abundance developed before the arrival of zebra mussels probably wouldn't apply after these invaders arrived. Problems with extending empirical models beyond their initial domains will be especially severe if the models are overfitted (see below).

As is the case with mechanistic models, there is an essentially infinite number of model structures for empirical models, and it may be difficult to choose among model structures (cf. Burnham and Anderson 2002). For instance, consider the relatively simple model in which we model mussel abundance as a function of phosphorus loading (a surrogate for primary production) to lakes. The link between phosphorus loads and mussel abundance could be modeled as linear, semilogarithmic, log-log, some sort of saturating function, unimodal (e.g., parabolic), a step-function or many other forms. Very frequently, we will not have sufficient prior knowledge or data to choose among these alternatives. The problem of choosing model structure becomes very much more severe when we consider several to several dozen variables, and may make complicated empirical models intractable, as is the case with complicated mechanistic models.

Paradoxically, the same abundance of environmental data that makes empirical analyses feasible and so attractive also poses technical problems for their design and interpretation. Consider the example shown in Table 12, in which data on unionoid abundance (expressed as catch-per-unit-effort in timed searches) and environmental variables were collected at 20 sites in southern New York. A conventional stepwise multiple linear regression analysis of the data produced a highly significant model, and identified several logical variables as good predictors of unionoid abundance (Table 13). So far, so good. Unfortunately, the analysis in Table 13 is completely spurious, because the "data" in Table 12 aren't real data, but literally a set of random numbers that I made up. Despite the complete absence of any real pattern in the data, our analysis was able to produce a plausible and "statistically significant" interpretation that could seriously mislead us. This problem, which occurs whenever the number of independent variables is large compared to the number of samples, is an increasingly frequent situation in this era of automated data collection.

Although this problem is well known among statisticians (e.g., Flack and Chang 1987, Anderson et al. 2001), flawed analyses like those of Tables 12 and 13 seem to be increasingly common in ecology as a result of the availability of large data sets that can be used as predictor variables. There are at least two ways to avoid making erroneous conclusions from analyses of data sets with a large number of independent variables. First, reduce the number of independent variables to a number that is reasonable compared with the number of samples. Sometimes one hears rough rules of thumb that the number of samples should be at least five or ten times the number of independent variables. The number of independent variables can be reduced by using methods such as principal components analysis (PCA) or by constructing synthetic variables with specific functional significance. The Palmer Drought Severity Index is an example of a synthetic variable that is used widely in agriculture and terrestrial ecology (Alley 1984); it combines information on soil characteristics and a series of past values of precipitation and to produce a single variable that is supposed to indicate drought severity. Probably the most straightforward way to reduce the number of independent variables, though, is to include only those variables that are genuinely thought to have strong effects on the dependent variable, or relate to the hypothesis that is being tested.

Second, one may use even data sets with a large number of independent variables as long as the interest is to *generate* rather than *test* hypotheses. An appropriate conclusion from Table 13, then, might be to design more focused studies to test whether mussel populations really are related

TABLE 12 *Mussel Population Density and Environmental Data at 20 Sites in New York Streams*

CPUE (no./min.)	0.93	0.73	0.74	0.90	0.92	0.75	0.04	0.76
Mean velocity	0.87	0.23	0.44	0.32	0.67	0.53	0.94	0.48
Maximum velocity	0.44	0.54	0.88	0.81	0.58	0.66	0.95	0.94
Minimum velocity	0.98	0.49	0.29	0.84	0.94	0.80	0.27	0.52
SD (velocity)	0.19	0.95	0.46	0.71	0.28	0.41	0.51	0.08
Mean tau	0.28	0.51	0.83	0.69	0.05	0.38	0.9	0.88
Maximum tau	0.39	0.72	0.98	0.57	0.03	0.86	0.97	0.79
Froude number	0.93	0.66	0.50	0.15	0.92	0.08	0.08	0.28
Mean temperature	0.18	0.72	0.10	0.68	0.41	0.43	0.15	0.02
Maximum temperature	0.89	0.50	0.11	0.28	0.14	0.43	0.59	0.11
Minimum temperature	0.83	0.56	0.58	0.89	0.71	0.32	0.15	0.01
SD (temperature)	0.14	0.91	0.76	0.10	0.75	0.80	0.95	0.20
pH	0.29	0.69	0.81	0.31	0.97	0.18	0.78	0.01
Mean dissolved oxygen	0.97	0.03	0.04	0.98	0.82	0.86	0.15	0.41
Minimum dissolved oxygen	0.92	0.65	0.07	0.47	0.29	0.53	0.25	0.52
Soluble reactive P	0.89	0.34	0.96	0.52	0.94	0.50	0.67	0.69
Dissolved inorganic N	0.81	0.09	0.81	0.41	0.15	0.96	0.89	0.41
Ammonium N	0.80	0.07	0.70	0.39	0.32	0.83	0.48	0.93
Dissolved organic C	0.33	0.48	0.74	0.73	0.47	0.71	0.85	0.95
Sediment phi	0.16	0.01	0.39	0.40	0.34	0.14	0.56	0.80
Embeddedness	0.58	0.62	0.79	0.91	0.70	0.01	0.46	0.66
Sediment organic C	0.25	0.08	0.89	0.65	0.34	0.54	0.94	0.61
Percent impervious surface in watershed	0.68	0.41	0.24	0.67	0.09	0.76	0.15	0.32
Road density in watershed	0.57	0.73	0.56	0.29	0.53	0.51	0.78	0.62
Mean watershed slope	0.58	0.58	0.55	0.65	0.80	0.82	0.08	0.89
Drainage density	0.04	0.52	0.61	0.11	1.00	0.98	0.77	0.12

NOTE: Variables are all standardized to lie between 0 and 1. Strayer and Lowe (unpublished).

to dissolved inorganic nitrogen, maximum temperature, or watershed slope, but not to claim that we have yet demonstrated any such relationships. Sometimes it is possible to perform these tests as part of the empirical analysis itself. For example, we might develop a statistical model using 75 study sites, and then test our model using 25 sites that we deliberately

TABLE 12 *Mussel Population Density and Environmental Data at 20 Sites in New York Streams (continued)*

0.40	0.73	0.11	0.31	0.82	0.09	0.68	0.06	0.53	0.17	0.81	0.46
0.58	0.08	0.21	0.58	0.40	0.28	0.60	0.22	0.34	0.72	0.95	0.34
0.72	0.82	0.88	0.14	0.66	0.18	0.34	0.81	0.07	0.47	0.05	0.09
0.03	0.58	0.46	0.72	0.14	0.47	0.48	0.28	0.48	0.94	0.57	0.10
0.24	0.14	0.77	0.39	0.25	0.99	0.91	0.03	0.95	1.00	0.79	0.31
0.35	0.47	0.21	0.19	0.22	0.26	0.16	0.72	0.13	0.93	0.59	0.33
0.68	0.42	0.66	0.37	0.10	0.62	0.67	0.07	0.99	0.30	0.74	0.49
0.07	0.07	0.38	0.29	0.82	0.58	0.92	0.57	0.10	0.66	0.13	0.37
0.19	0.32	0.07	0.90	0.25	0.03	0.59	0.10	0.69	0.95	0.95	0.92
0.28	0.32	0.01	0.16	0.91	0.89	0.88	0.11	0.36	0.01	0.96	0.89
0.46	0.68	0.97	0.33	0.62	0.33	0.41	0.60	0.01	0.52	0.16	0.44
0.28	0.41	0.33	0.10	0.60	0.23	0.31	0.86	0.75	0.81	0.14	0.66
0.38	0.20	0.16	0.17	0.29	0.41	0.74	0.45	0.84	0.15	0.57	0.54
0.90	0.96	0.90	0.36	0.95	0.92	0.08	0.12	0.38	0.86	0.96	0.46
0.61	0.29	0.79	0.29	0.27	0.38	0.11	0.86	0.26	0.55	0.08	0.58
0.98	0.51	0.88	0.35	0.73	0.15	0.06	0.29	0.69	0.83	0.90	0.16
0.50	0.60	0.48	0.26	0.86	0.65	0.27	0.22	0.74	0.06	0.35	0.62
0.92	0.99	0.52	0.60	0.04	0.73	0.81	0.45	0.25	0.99	0.20	0.31
0.46	0.71	0.78	0.96	0.78	0.42	0.96	0.30	0.81	0.47	0.01	0.11
0.59	0.13	0.44	0.35	0.74	0.44	0.76	0.39	0.76	0.61	0.36	0.92
0.03	0.79	0.27	0.87	0.44	0.72	0.39	0.45	0.71	0.16	0.88	0.13
0.01	0.63	0.37	0.17	0.11	0.48	0.67	0.67	0.16	0.45	0.14	0.20
0.59	0.29	0.03	0.10	0.64	0.73	0.80	0.56	0.22	0.18	0.11	0.18
0.65	0.89	0.99	0.93	0.46	0.22	0.27	0.16	0.96	0.20	0.64	0.86
0.15	0.79	0.08	0.47	0.88	0.45	0.94	0.95	0.00	0.95	0.04	0.87
0.67	0.46	0.73	0.32	0.26	0.49	0.70	0.96	0.42	0.96	0.23	0.18

held back from the initial analysis. A special case of this approach is the jackknife (Efron and Tibshirani 1993), which may also be useful.

Finally, I note that interactions among controlling variables are ignored in empirical models just as often as in mechanistic models, although there is no reason why interactions cannot be included in empirical models. If

TABLE 13 *Results of a Stepwise Multiple Linear Regression Analysis of the Data in Table 12*

Variable	F	p
Maximum current	3.6	0.09
Minimum current	3.9	0.08
Mean shear stress	10.2	0.01
Maximum shear stress	12.1	0.007
Froude number	7.0	0.03
Maximum temperature	29.9	<0.001
Dissolved inorganic N	48.7	<0.001
Percent embeddedness	16.5	0.003
Mean watershed slope	31.3	<0.001
Drainage density	4.8	0.06

NOTE: $p = 0.15$ to enter or remove variables. The final model has $R^2 = 0.93$ and $p = 0.0006$. Strayer and Lowe (unpublished).

the number of independent variables is large, however, the number of interactions will be too large to be easily estimated using empirical approaches. For instance, the number of two-way interactions among n independent variables alone is

$$\frac{n(n-1)}{2}.$$

Several empirical analyses attempting to relate environmental variables (i.e., habitat in the broad sense) to mussel presence, species richness, or abundance have already been published (Table 14). The number of independent variables in most of these studies is high compared to the number of sites studied. Furthermore, none of the studies published so far has considered the problem of spatial autocorrelation among sites, which may reduce the number of effectively independent sites even further (see Fortin and Dale (2005) for a discussion of this problem and possible solutions). Therefore, most of the empirical studies of mussels published so far must be considered exploratory rather than definitive, and will need to be followed up by more focused or confirmatory studies. Notably, only Arbuckle and Downing (2002) tried to confirm or test their findings with independent data sets, and they were able to find only weak confirmation of their initial findings. The studies published so far have either had low predictive power or a very low number of sites compared to the number of independent variables, suggesting that the reported predictive power may have been inflated by spurious relationships. Finally, none of the published models seems to have included interactions among independent variables. Thus, we have some distance to go before we have empirical models for mussel populations that have been tested and have satisfactorily high predictive power.

Despite these potential problems, I think that careful empirical analyses have good potential for predicting the distribution and abundance of unionoids from multiple controlling variables. The flexibility of empirical models, the wide availability of statistical software, and the rapidly increasing availability of environmental data at multiple scales make empirical modeling attractive. Problems are likely to arise if the model includes a large number of independent variables relative to the number of sites studied, if those variables are chosen without a reasonable expectation that they are related to mussel populations, or if models developed for one set of circumstances are thoughtlessly extrapolated to other sites or times. Finally, I suspect that the limited availability of reliable and comparable data on mussel populations will set the ultimate limit on the contributions of empirical models in mussel ecology. At present, we have relatively few data sets on mussel populations suitable for empirical modeling, especially in comparison to the very large data sets on environmental data. Empirical models will not reach their full potential until we systematically collect many more data on mussel populations.

TABLE 14 *Summary of Empirical Analyses Relating Mussel Populations or Communities at the Stream Reach Scale to Characteristics of the Site or its Watershed*

Location	Dependent variables	Number of sites	Number of independent variables	% of variance accounted for (R^2 or ASCC)	Source
Iowa	Species richness, density	118	17	27–36	Arbuckle and Downing (2002)
Georgia	Species richness, density, diversity	46	7–18	16–49	Gagnon et al. (2006)
Michigan	Abundance of all mussels; abundance of seven common species	40	40	51–86	McRae et al. (2004)
Iowa	Loss in species richness	118	16–17	23–51	Poole and Downing (2004)
Washington	Presence/absence of *Margaritifera falcata*	31	17[a]	67	Stone et al. (2004)
Mid-Atlantic	Presence/absence of 13 species	141	6	3–55	Strayer (1993)

[a] Less an unknown number of variables that were excluded from the analysis because of a high variance inflation factor.

TEN

IS A COMPREHENSIVE MODEL POSSIBLE?

I do not believe that any of the three popular approaches to integration, as currently practiced by ecologists, is likely to lead to a satisfactory predictive understanding of unionoid distribution and abundance. As I suggested earlier (and will discuss further below), it is likely that a simple Liebigian approach will be satisfactory only when applied over limited domains. Uncritical attempts to apply a single-factor approach over a broad range of conditions may well lead to fruitless arguments about which factor is "the" limiting factor, when in reality multiple factors probably limit different mussel populations. Certainly, ecology has seen many such attempts to reduce a multifactorial world to a single factor, leading to arguments about what that master factor is (e.g., the top-down vs. bottom-up controversy in plankton ecology; whether predation, competition, or disturbance is *the* master factor that controls community structure; whether communities are controlled by local *or* regional processes, etc.). These arguments probably have generated more heat than light.

I believe that a fully mechanistic, multifactor model probably is the implicit ultimate goal of many ecologists. However, I do not think we have the abilities to collect data or to produce and test such models, and that it is counterproductive to conduct our research as if we were contributing to the construction of such a chimera. In the particular case of unionoid ecology, we have even fewer data in hand, fewer workers, and fewer resources than for other research problems where this approach has been rejected as unworkable (e.g., Rigler and Peters, 1995).

Although I think that empirical models have promise to advance unionoid ecology, they are at present a long way from fulfilling that promise. Data sets are often so small that they limit the exploration of the influence of multiple factors to the most preliminary of analyses rather than allowing reliable model-building, and technical statistical problems with model definition and selection may be difficult to solve. There has been almost no attention given to the critical issue of defining the domain over which each empirical model applies or to testing the conclusions reached from an empirical model with independent evidence (the work of Arbuckle and Downing 2002 being a notable exception). Finally, empirical models will inevitably fail to satisfy scientists who are chiefly concerned with mechanistic understanding.

How, then, do we proceed? Must we abandon the problem of predicting unionoid distribution and abundance as unmanageably complex? I do not claim to be able to describe precisely the models and research that will lead to a solution of this problem, but I think it might be possible to outline promising research strategies that might be used. Specifically, I suggest that a successful research program probably will contain four key elements: (1) clear definition of goals and needs; (2) regular and rigorous testing of models; (3) addition of complexity deliberately and only as needed; and (4) production of a series of models with limited domains instead of a Grand Unifying Theory.

First, I think it is useful to define our research goals clearly. A good goal can be used to focus and evaluate research efforts. A goal like "to understand the factors that control unionoid communities" seems to me to be too vague to be really useful. What aspect of unionoid communities: species richness, relative abundance of species, distribution, abundance, or viability of individual species? Likewise, unless we have some idea what we mean by "understand" or how much understanding we require, it will be difficult to know if we've reached our goal. I suggest that predictive power often is a reasonable goal, because the ability to predict often is useful in applied settings and has the advantage of being measurable. Further, as I have already suggested, perfect understanding or predictive power is not a reasonable or necessary goal, so the process of defining goals ought to make us think about how much predictive power we really need.

So as a first step, I suggest that individual researchers or managers, or the field as a whole, need to devote some thought to precisely what the goals of our research are. I am not suggesting that it is necessary for all of us to agree on a single goal (although our lives would be much easier if we could!), because different people have different interests and management needs, but rather that each of us be able to clearly state and defend a goal of his or her research program. As an example, personally I would like to be able to predict the local (1-km reach) abundance of unionoid species within a factor of three. I would also like to be able to predict the viability (e.g., the probability that a given unionoid population will survive for the next 30 years) of such unionoid populations, but I am not sure that we are close enough to being able to either make or test such predictions to make this a reasonable goal.

Second, I think it is critical that we subject our ideas and models to regular and rigorous tests. Much of the power of science comes from the routine and careful testing of ideas, which leads to rapid refinement of ideas. If we don't test our ideas and models, we can't refine them. Unfortunately, relatively few models in unionoid ecology have been subject to rigorous, independent tests, especially if we're thinking about specific, parameterized models. This deficiency may stem from inadequate research funding, a feeling among academics and funders that "repeating" a research study is derivative and not intellectually compelling, and the absence of a culture of actual model-testing among unionoid ecologists. We do not tolerate inadequate testing of ideas and models in other fields that matter to us (e.g., biomedical research), and if we think that ecological knowledge is important, then we must routinely test and refine that knowledge. Whatever the reasons for inadequate testing, I think that progress in unionoid ecology will be slow until we develop the habit of regularly funding and carrying out independent tests of our important models and ideas.

Third, I think we need to begin with simple models and add complexity deliberately and only as needed. Adding complexity vastly increases both the amount of data we need to collect and the difficulty of building and testing models. This is especially true if we include interactions among variables, which greatly increases the number of parameters that must be estimated. I would argue, therefore, that adding complexity will substantially delay the construction and testing of models, and is more likely to slow progress than speed it. I suggest that complexity be added only after simple models are positively shown to fail to reach our goals, and that once added, the benefits and costs of the added complexity be evaluated carefully.

Finally, I think that at least initially it will be more useful to think about building a series of small models with limited domains than to try to build a grand model that applies to all habitats and unionoids. Although I have argued that all five classes of factors—dispersal, habitat, fish hosts, food, and predators—are important in some situations, they probably are not all important in all situations. Thus, it should be possible to build satisfactory models with fewer than five factors that apply to a limited range of situations. Such reduced models should be much easier to build and test than a grand model, just as a regional identification key to a small number of local species is easier to use and more reliable than a global key to the unionoids of the world. For instance, we might try to build and test a model that applies just to a single ecoregion or to a particular habitat type. The strategy of building models with limited domains will require careful attention to defining (and testing) the intended domains of the model.

MY FAVORITE MONSTERS (1)

Because I have been so free with advice, I feel some responsibility to offer a specific model or two for discussion. I am not suggesting that these are actually the specific models that will solve the problems of unionoid ecology—we have not yet adequately assessed the importance of several key processes that might be important (see the list on pp. 117-118), and it seems arrogant to suppose that I can conjure the right model from the comfort of my office. Nevertheless, I offer these models here merely as concrete examples of how one might build models of unionoid distribution and abundance, and hope they will inspire readers to produce their own, better, models.

First, suppose we decide that we want to predict the presence/absence or population density of a mussel species at a spatial scale of 1-km long reaches of stream. It seems efficient to use a hierarchical approach like that suggested by Guisan and Thuiller (2005) (Fig. 49). I will suppose that broad-scale dispersal determines whether the species occurs in the region at all, and that we have ample distributional data to determine that the species lives in our study region.

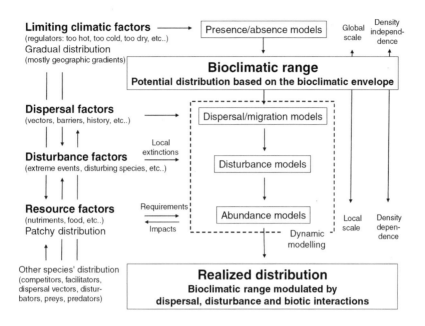

FIGURE 49. A general hierarchical model for predicting the distribution or abundance of a species, from Guisan and Thuiller (2005).

Second, I will assume that the extent of suitable habitat determines the proportion of each 1-km reach that is suitable for mussels. I focus first on habitat because I think habitat quality constrains many unionoid populations, because we know quite a lot about habitat requirements of unionoids, and because it is relatively straightforward to collect relevant information about habitat quality.

As discussed earlier, several attributes of the habitat appear to be important to unionoids. The entire reach might be scored as suitable or unsuitable based on its temperature regime and water chemistry. If the thermal regime and water chemistry are suitable, the extent of suitable habitat might be defined as areas of the stream bottom that are under water during droughts (e.g., a 7-day, 10-year drought) but not scoured during high water (again, e.g., a 10-year flood), that contain sediment soft enough for burrowing but firm enough to support mussels, and that do not contain excessive free ammonia during summer or contain excessive toxins. This could require application of either detailed (e.g., Howard and Cuffey 2003, Morales et al. 2006) or simple (Lamouroux et al. 1992) hydraulic models, along with field surveys of sediment penetrability, interstitial free ammonia, and other toxins. The results of this step might look like Fig. 50, providing a value for habitat quality for each reach in terms of the area (or proportion) of stream bottom that is suitable habitat. Although I have said that it is straightforward to collect habitat information, it would be a lot of work to collect enough data to produce maps like Fig. 50 for a large number (e.g., 20-50) of study reaches.

I think it is unlikely that models based solely on habitat will provide satisfactory predictive power. I note in particular that the distributions of different unionoid species often are congruent within reaches (Fig. 51), although the different species typically have very different abundances, and species distributions are usually nested across reaches (Fig. 52). For this pattern to arise from habitat alone, it would require that different unionoid species have very different habitat breadths (Fig. 53), in terms of the physical and chemical factors that they tolerate. Although there is some evidence that different unionoid species have different tolerances for physicochemical variables, these differences are much subtler than would be needed to account for the patterns in Figs. 51 and 52. Instead, the patterns in Figs. 51 and 52 suggest to me that some other controlling factor whose influence differs substantially among species is overlaid on habitat needs that are similar across most unionoid species. (This infor-

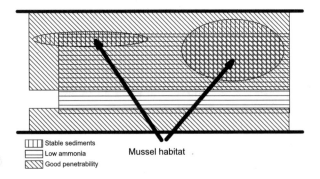

FIGURE 50. A map of the extent of suitable habitat in a fictional reach of stream. Suitable habitat includes areas with stable sediments, low ammonia, and moderate penetrability.

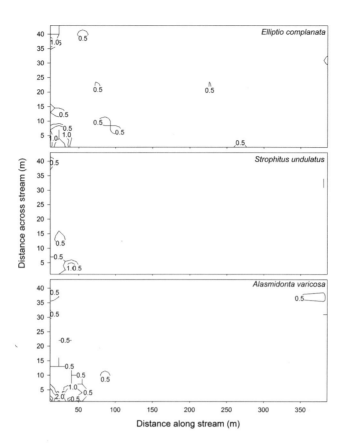

FIGURE 51. Example showing that different mussel species often have similar spatial distributions within a stream reach. Contour plots show the densities (number/m²) of the three most abundant mussel species in a section of the Neversink River, New York. From unpublished data associated with the study of Strayer (1999a).

mal assessment of the adequacy of models based solely on habitat should be supplanted by a rigorous test, once we have the necessary data).

I will guess that this additional factor is host availability, so that population density within habitat patches is controlled by the availability of fish hosts. I make this assumption because we have evidence of highly differential host breadth among unionoid species (e.g., Figs. 29, 30, Tables 6, 7), so that interspecific differences in host availability could be large enough to account for patterns like those in Figs. 51 and 52. Further, host fish abundances vary widely within and across reaches that support mussel

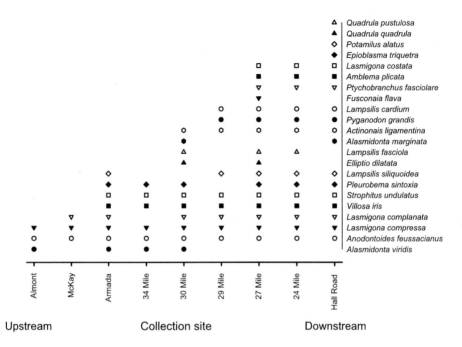

FIGURE 52. An example of nested species distributions across reaches, as is often observed in mussel communities. Symbols show the occurrence of mussel species along the North Branch of the Clinton River, Michigan, in 1978. Sites are listed in order from upstream (left) to downstream (right). From data of Strayer (1980).

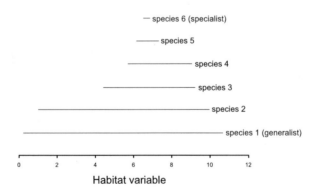

FIGURE 53. Hypothetical diagram of nested habitat requirements of a group of coexisting unionoid species. Lines show the range of physicochemical conditions tolerated by each species. Nested habitat requirements such as these would be required to account for patterns like those shown in Figs. 51 and 52 through habitat alone.

populations. We do not as yet have comparable evidence of such differential influences of food or enemies on different unionoid species.

To incorporate host availability into a predictive model of unionoid distribution and abundance, we would obviously need to know which fish species served as hosts for the mussel species of interest. To simplify matters, we might choose to ignore fish species that either were marginal hosts in laboratory studies or were uncommon at our study sites. Then we would need field data on the abundance of hosts at our study sites. Ideally, we would do fish surveys over the areas previously identified as suitable mussel habitat, during the season of glochidial release (to best estimate contact probabilities between glochidia and hosts), but we might have to settle for host densities on a reach-wide scale at other seasons. Because of the evidence that older, larger fish often serve as poor hosts (see Chapter 5 on fish hosts for details), we might choose to collect the fish data by size-class.

If the mussel species uses more than one fish species as a host, it is likely that each host species carries different numbers of glochidia, as described by Martel and Lauzon-Guay (2005). The glochidial load carried by each species is almost certainly most efficiently estimated from field data. Although Martel and Lauzon-Guay (2005) found substantial variation across sites in the relative glochidial load carried by different species, I hope it will not be necessary to measure glochidial loads at all sites. Instead, I hope it will be sufficient to use an average glochidal load for each host species measured at a few sites to provide an index of the capacity of different host species. The total host availability at each site would therefore be

$$\sum_{ij} c_{ij} F_{ij}$$

where c_{ij} is the mean number of glochidia carried by the ith size-class of the jth host species, and F_{ij} is the number of fish of that size and species at the site.

From this point on, the model probably is best evaluated purely empirically to avoid having to parameterize a model like those discussed in Chapter 5 on host availability. Instead, we could model the probability of occurrence or abundance of a mussel species in each reach as a function of the amount or proportion of suitable habitat and host availability. If the equations from the section on fish-immunity models are a reliable guide, I do not expect mussel occurrence and population density to be linear functions of the extent of suitable habitat or host availability (Fig. 35).

Such a model is both objectionably simple and impractically difficult to evaluate in practice. As this example shows, any attempt to develop quasi-mechanistic models of mussel distribution and abundance must navigate between the Scylla of oversimplicity and the Charybdis of intractable complexity. Despite the apparent complexity of this particular model, note that I have deliberately left out many key processes; for example, predation; food-limitation; the likelihood that the probability of glochidium-host contact varies across sites and years; the within-reach distribution of settling juveniles; the role of local dispersal in linking metapopulations or in creating sink populations; the whole issue of temporal variability; the possibility that habitat quality affects the density as well as the presence of mussels; and the possibility of Allee effects driving sparse populations to extinction. These might be added if the predictive power of the basic model is inadequate and we are able to collect the data and successfully model the increased complexity.

Once the data are collected, we must test to see whether the model satisfactorily predicts unionoid abundance from habitat and fish host data. If the model's performance meets our goals (e.g., predicting abundance within a factor of three), then we are finished and can offer up the model for others to test and use in other situations. In the more likely event that the initial model fails to meet our goals, we will need to look closely at the pattern of model failure (e.g., does the model work for small streams but not large rivers? does it work for host specialists but not host generalists?) and consider possible refinements of the model (e.g., changing the way in which we calculated habitat availability or host availability or adding additional variables such as muskrat predation).

I have described this model in some detail as a straw man to show how one might go through the process of building a quasi-mechanistic model, and to convince you that building even a simple model that omits many details is an ambitious task. I am sure that there are other relatively simple models that are worth exploring, but without doubt it is worthwhile to lay out your favorite model to evaluate its feasibility and identify exactly what data are needed before going into the field.

MY FAVORITE MONSTERS (2)

Again for the purpose of discussion, I offer another example of a possible model, more empirical than the first. Again, assume that we are interested in predicting presence/absence or population density of mussels at the scale of 1-km long stream reaches, and that we want to test the hypothesis that recent widespread declines in mussel populations are a result of ammonia toxicity that kills juveniles (cf. Augspurger et al. 2003, Newton 2003, Mummert et al. 2003). There are some data on trends in unionoid populations at sites across North America since the 1970s or 1980s that could be used for this model (e.g., Ahlstedt and Tuberville 1997, Metcalfe-Smith et al. 1998, 2003, Strayer and Fetterman 1999, Fraley and Ahlstedt 2000, Vaughn 2000, Poole and Downing 2004, Warren and Haag 2005, and other unpublished or grey literature studies). As a response variable, we might simply score sites as having lost or retained unionoid populations between 1970–1980 and 1995–2005. We might also score populations at sites that retained unionoids as recruiting (juveniles present in 1995–2005) or not recruiting (juveniles absent), if this information is available for enough sites. It might seem most straightforward to see if measured concentrations of unionized ammonia in stream sediments exceed values known to kill juvenile mussels in laboratory tests, and test if streams with excessive unionized ammonia have failing unionoid populations. However, I do not think that suitably extensive data on interstitial ammonia concentrations exist.

Rather than abandoning this line of inquiry, we might design an empirical study to look for evidence of widespread ammonia toxicity. We might expect that levels of interstitial unionized ammonia should depend on nitrogen availability, system productivity, pH, temperature, and interstitial water flow. Thus, we could collect information on concentrations of inorganic nitrogen (nitrate plus ammonia), phosphorus (which often limits productivity in freshwater ecosystems), pH, and maximum temperature in the water column in the 1980s and 1990s. Such data are available for many sites in the United States and Canada in publicly available databases collected by national and state programs to monitor water quality, as well as from individual research programs (e.g., EPA 2006, NAWQA 2006, Ohio EPA 2006, USGS 2006a, b). We might use stream slope or even local landscape slope (from digital elevation maps or USGS quadrangles) as a rough indicator of interstitial water flow, reasoning that streams with high slopes or in steep landscapes usually have coarse, permeable sediments and high hydrologic gradients.

Although there are many ways to analyze such data, logistic regression might be the simplest. Our hypothesis is that failing unionoid populations would be associated with a combination of high concentrations of inorganic nitrogen and phosphorus, high pH and temperature, and low stream slopes. Several aspects of the model output are useful. Obviously, we are interested in the overall predictive power of the model: does the model accurately predict the status of unionoid populations across the study sites? Does it meet the goals that motivated our study? Second, we are of course interested in the signs, regression coefficients, and errors associated with each of the independent variables. Third, it is useful to examine carefully the residuals from the model predictions to look for geographic areas or habitats in which the model is prone to fail, and to get clues as to how we might improve the model or extend its domain.

As is the case with mechanistic models, empirical modeling is an iterative process. Once we see the results of the initial model, we can adjust its structure, add complexity, or restrict its predictive domain. Model results can also be very helpful in designing subsequent field campaigns or experiments. For instance, our model results might help us decide where it would be most productive to actually measure interstitial ammonia.

CAN WE MAKE THE MONSTER WALK?

I return finally to the central question of this essay: is it feasible to produce a quantitative theory for predicting the distribution and abundance of unionoid mussels? We will not finally know the answer to this question until scientists have seriously tried to produce and test such theories, and have failed repeatedly or succeeded. Such attempts have not yet been made. Nevertheless, I see no reason at this point to conclude that this is an intractable scientific problem. As I have suggested above, I believe that the construction of a comprehensive, mechanistic model is a hopeless enterprise. Some readers may disagree with or be disappointed by this conclusion. On the other hand, I think there are several promising alternatives to such comprehensive models that can be pursued, and which may give satisfactory results. Whatever course we choose, it is clear that we will need more insight into key processes that control unionoid populations, some sort of integrative framework, and deliberate, rigorous tests of our theories.

PATHWAYS TO PROGRESS?

How do we best proceed toward the goal of predicting unionoid distribution and abundance? To begin with, we are going to need more information to build and test integrative models of unionoid distribution and abundance, regardless of the modeling approaches that we use. In addition to the specific research needs that I identified above, I think we will need better information on the status of unionoid populations. Although we are collecting more and better information on unionoid populations each year, our analyses are still severely limited by inadequate data. Reliable, quantitative data on unionoid populations are still rare, and it is even rarer for those data to be collected at the large spatial scales (i.e., more than a single mussel bed) at which controlling factors operate, and at which we might want to build our models. Rarer still are data on the demographic status of unionoid populations, including survivorship and fecundity schedules. Such demographic data have the potential of serving as leading indicators of the status of unionoid populations, and may allow us to move beyond the analysis of the effects of past events on unionoid populations to the interpretation of the present status and future prospects of mussel populations.

It is also unfortunate that there are so few long-term data on unionoid populations. Long-term data would seem to be essential to understand the demography of animals with such long life spans, especially considering the enormous temporal variability in the streams and rivers that are the primary habitat of many of these animals. It is therefore remarkable that there are almost no published studies that track a unionoid population for even 10 years (I am aware of only a few examples: Lewandowski 1991, Payne and Miller 2000, Schloesser et al. 2006, Strayer and Malcom 2007a, some of which describe pathologically declining communities). Carefully designed long term studies of a few unionoid communities could be very valuable. Such quantitative, spatially extensive, demographically detailed, or long term data on unionoid populations will not always be inexpensive or easy to collect, but I think it will be essential for producing and testing models of mussel distribution and abundance.

Just as important, I believe that we should devote more thought to matching the data we collect with the models that will ultimately use those data. We should collect data because we actually need it for something, not just because no one has ever collected those data before. Thus, I believe that it would be useful if we all thought a little harder about how the information we are collecting actually could be used to answer the big questions that motivate us, as we toil in our scientific laboratories or graveyards.

I suspect that one result of this critical introspection will be more sympathy for empirical approaches. In any case, I think that careful reflection about ways individual studies can be fitted together should result in a body of information that is better suited to our ultimate purposes.

Finally, in setting down my own thoughts about research goals, approaches, and priorities concerning freshwater mussels, I couldn't help but wonder whether the larger community of people interested in freshwater mussel ecology and management has optimized the procedure for identifying and focusing resources on the important research and management needs. To the extent that the field has any direction, it is set by a thousand more or less uncoordinated decisions about what research to attempt, what management actions to implement, what papers and grant proposals to accept or reject, and so on. The assumption of this approach is that the best ideas will succeed in this marketplace of ideas, thereby optimizing the impact of our limited resources. There are other approaches for setting priorities, however. Influential books and reviews (e.g., Dillon 2000, Vaughn and Hakenkamp 2001) certainly can help to set research priorities, and can offer broad coverage of the field, but ultimately represent the opinions of one or a few authors rather than the entire community. The "marketplace of ideas" approach sometimes is supplemented by special sessions at scientific meetings or ad hoc task forces that highlight promising research directions or pressing management needs. Meetings or workshops are even designed specifically to identify critical research challenges, management needs, or promising approaches (e.g., Carpenter 1988, Naiman et al. 1995, National Native Mussel Conservation Committee 1998). However, the recommendations of such workshops often are broad or unprioritized (presumably to avoid offending any of the report's diverse authors or readers), and I do not know how much such reports actually cause researchers, funders, or managers to change the direction of their activities. Consequently, it's not clear to me how effective they are in guiding research or management activities. Can concerted, community-wide efforts successfully identify and put priorities on critical research and management activities, or are they politically infeasible? Worse yet, are they less efficient than a "marketplace of ideas" approach in steering research and management? It may be worth devoting some thought to these questions because of their potential to improve our abilities to focus our limited resources more effectively on our ultimate goals of better understanding and management of freshwater mussels.

LITERATURE CITED

Aber, J.D., C.L. Goodale, S.V. Ollinger, M.L. Smith, A.H. Magill, M.E. Martin, R.A. Hallett, and J.L. Stoddard. 2003. Is nitrogen deposition altering the nitrogen status of northeastern forests? BioScience 53: 375–389.

Aber, J., W. McDowell, K. Nadelhoffer, A. Magill, G. Berntson, M. Kamakea, S. McNulty, W. Currie, L. Rustad, and I. Fernandez. 1998. Nitrogen saturation in temperate forest ecosystems: Hypotheses revisited. BioScience 48: 921–934.

Ackerman, J.D., M.R. Loewen, and P.F. Hamblin. 2001. Benthic-pelagic coupling over a zebra mussel reef in western Lake Erie. Limnology and Oceanography 46: 892–904.

Adams, C.C. 1892. Mollusks as cat-fish food. The Nautilus 5: 127–128.

Ahlstedt, S.A., and J.D. Tuberville. 1997. Quantitative reassessment of the freshwater mussel fauna in the Clinch and Powell Rivers, Tennessee and Virginia. Pages 72–97 *In*: K.S. Cummings, A.C. Buchanan, C.A. Mayer, and T.J. Naimo (eds.). Conservation and management of freshwater mussels. II. Initiatives for the future. Upper Mississippi River Conservation Committee, Rock Island, IL. 293 pp.

Aldridge, D.C., and A.L. McIvor. 2003. Gill evacuation and release of glochidia by *Unio pictorum* and *Unio tumidus* (Bivalvia: Unionidae) under thermal and hypoxic stress. Journal of Molluscan Studies 69: 55–59.

Allan, J.D., R. Abell, Z. Hogan, C. Revenga, B.W. Taylor, R.L. Wellcome, and K. Winemiller. 2005. Overfishing of inland waters. BioScience 55: 1041–1051.

Allen, W.R. 1914. The food and feeding habits of freshwater mussels. Biological Bulletin 27: 127–147.

Alley, W.M. 1984. The Palmer Drought Severity Index: Limitations and assumptions. Journal of Climate and Applied Meteorology 23: 1100–1109.

Amyot, J.-P., and J.A. Downing. 1998. Locomotion in *Elliptio complanata* (Bivalvia: Unionidae): A reproductive function? Freshwater Biology 39: 351–358.

Anderson, D.R., K.P. Burnham, W.R. Gould, and S. Cherry. 2001. Concerns about finding effects that are actually spurious. Wildlife Society Bulletin 29: 311–316.

Anthony, J.L., and J.A. Downing. 2001. Exploitation trajectory of a declining fauna: A century of freshwater mussel fisheries in North America. Canadian Journal of Fisheries and Aquatic Sciences 58: 2071–2090.

Anthony, J.L., D.H. Kesler, W.L. Downing, and J.A. Downing. 2001. Length-specific growth rates in freshwater mussels (Bivalvia: Unionidae): Extreme longevity or generalized growth cessation? Freshwater Biology 46: 1349–1359.

Araujo, R., and M.A. Ramos. 2000. Status and conservation of the relict giant European freshwater pearl mussel *Margaritifera auricularia* (Spengler, 1793) (Bivalvia: Unionoidea). Biological Conservation 96: 233–239.

Arbuckle, K.E., and J.A. Downing. 2002. Freshwater mussel abundance and species richness: GIS relationships with watershed land use and geology. Canadian Journal of Fisheries and Aquatic Sciences 59: 310–316.

Areekijseree, M., A. Engkagul, S. Kovitvadhi, U. Kovitvadhi, A. Thongpan, and K. Torrissen-Rungruangsak. 2006. Development of digestive enzymes and in vitro digestibility of different species of phytoplankton for culture of early juveniles of the freshwater pearl mussel, *Hyriopsis* (*Hyriopsis*) *bialatus* Simpson, 1900. Invertebrate Reproduction and Development 49: 255–262.

Arey, L.B. 1921. An experimental study on glochidia and the factors underlying encystment. Journal of Experimental Zoology 33: 463–499.

Arey, L.B. 1923. Observations on an acquired immunity to a metazoan parasite. Journal of Experimental Zoology 38: 377–381.

Arey, L.B. 1932. A microscopical study of glochidial immunity. Journal of Morphology 53: 367–379.

Arnott, D.L., and M.J. Vanni. 1996. Nitrogen and phosphorus recycling by the zebra mussel (*Dreissena polymorpha*) in the western basin of Lake Erie. Canadian Journal of Fisheries and Aquatic Sciences 53: 646–659.

Arter, H.E. 1989. Effect of eutrophication on species composition and growth of freshwater mussels (Mollusca, Unionidae) in Lake Hallwil (Aargau, Switzerland). Aquatic Sciences 51: 87–99.

Augspurger, T., A.E. Keller, M.C. Black, W.G. Cope, and F.J. Dwyer. 2003. Water quality guidance for protection of freshwater mussels (Unionidae) from ammonia exposure. Environmental Toxicology and Chemistry 22: 2569–2575.

Bailey, A. 1891. Shells of Erie Canal. The Nautilus 5: 23.

Baines, S.B., N.S. Fisher, and J.J. Cole. 2005. Uptake of dissolved organic matter (DOM) and its importance to metabolic requirements of the zebra mussel, *Dreissena polymorpha*. Limnology and Oceanography 50: 36–47.

Baird, M.S. 2000. Life history of the spectaclecase, *Cumberlandia monodonta* Say, 1829 (Bivalvia, Unionoidea, Margaritiferidae). M.S. thesis, Southwest Missouri State University, Springfield, IL. 108 pp.

Baker, A.M., F. Sheldon, J. Somerville, K.F. Walker, and J.M. Hughes. 2004. Mitochondrial DNA phylogenetic structuring suggests similarity between two morphologically plastic genera of Australian freshwater mussels (Unionoida: Hyriidae). Molecular Phylogenetics and Evolution 32: 902–912.

Baker, F.C. 1916. The relation of mollusks to fish in Oneida Lake. New York State College of Forestry and Syracuse University Technical Publication 4:1–366, Syracuse, NY.

Baker, F.C. 1922. The molluscan fauna of the Big Vermilion River, Illinois, with special reference to its modification as the result of pollution by sewage and manufacturing wastes. Illinois Biological Monographs 7:105–224.

Baker, F.C. 1926. The naiad fauna of the Rock River system: A study of the law of stream distribution. Transactions of the Illinois State Academy of Science 19: 103–112.

Baker, F.C. 1928. The fresh-water Mollusca of Wisconsin. Part II. Pelecypoda. Bulletin of the Wisconsin Geological and Natural History Survey 70: 1–495.

Baker, S.M., and J.S. Levinton. 2003. Selective feeding by three native North American freshwater mussels implies food competition with zebra mussels. Hydrobiologia 505: 97–105.

Baldigo, B.P., G.E. Schuler, and K. Riva-Murray. 2002. Mussel community composition in relation to macrohabitat, water quality, and impoundments in the Neversink River, New York. United States Geological Survey Open-File Report 02–104. 34 pp. Available at http://ny.water.usgs.gov/pubs/of/of02104/of02–104.pdf.

Baldwin, N.A., R.W. Saalfeld, M.R. Dochoda, H.J. Buettner, and R.L. Eshenroder. 2007. Commercial fish production in the Great Lakes 1867-2000. www.glfc.org/databases/commercial/commerc.php. Accessed 23 October 2007.

Balfour, D.L., and L.A. Smock. 1995. Distribution, age structure, and movements of the freshwater mussel *Elliptio complanata* (Mollusca: Unionidae) in a headwater stream. Journal of Freshwater Ecology 10: 255–268.

Barfield, M, and G.T. Watters. 1998. Non-parasitic life cycle in the green floater, *Lasmigona subviridis* (Conrad, 1835). Triannual Unionid Report 16: 22.

Bärlocher, F., and H. Brendelberger. 2004. Clearance of aquatic hyphomycete spores by a benthic suspension feeder. Limnology and Oceanography 49: 2292–2296.

Barnhart, M.C. 2006. *Epioblasma*—"fish snappers". http://unionid.missouristate.edu/gallery/Epioblasma/default.htm. Accessed 24 August 2006.

Barnhart, M.C., F. Riusech, and M. Baird. 1998. Hosts of salamander mussel (*Simpsonaias ambigua*) and snuffbox (*Epioblasma triquetra*) from the Meramec River system, Missouri. Triannual Unionid Report 16: 34.

Bartholomew, J.L., and J.C. Wilson (eds.). 2002. Whirling disease: Reviews and current topics. American Fisheries Society Symposium 29: 262 pp.

Bates, J.M. 1962. The impact of impoundment on the mussel fauna of Kentucky Reservoir, Tennessee River. American Midland Naturalist 68: 232–236.

Bauer, G. 1983. Age structure, age–specific mortality rates and population trend of the fresh–water pearl mussel (*Margaritifera margaritifera*) in North Bavaria. Archiv für Hydrobiologie 98: 523–532.

Bauer, G. 1987a. Reproductive strategy of the fresh-water pearl mussel *Margaritifera margaritifera*. Journal of Animal Ecology 56: 691–704.

Bauer, G. 1987b. The parasitic stage of the freshwater pearl mussel (*Margaritifera margaritifera* L.) II. Susceptibility of brown trout. Archiv für Hydrobiologie Supplementband 76: 403–412.

Bauer, G. 1987c. The parasitic stage of the freshwater pearl mussel (*Margaritifera margaritifera* L.) III. Host relationships. Archiv für Hydrobiologie Supplementband 76: 413–423.

Bauer, G. 1988. Threats to the freshwater pearl mussel *Margaritifera margaritifera* L. in central Europe. Biological Conservation 45: 239–253.

Bauer, G. 1991a. Spatial distribution of freshwater mussels: The role of host fish and metabolic rate. Freshwater Biology 26: 377–386.

Bauer, G. 1991b. Plasticity in life history traits of the freshwater pearl mussel: Consequences for the danger of extinction and for conservation measures. Pages 103–120 *in*: A. Seitz and V. Loeschke (eds.). Species conservation: A population-biological approach. Birkhäuser Verlag, Basel.

Bauer, G. 1992. Variation in the life span and size of the freshwater pearl mussel. Journal of Animal Ecology 61: 425–436.

Bauer, G. 1994. The adaptive value of offspring size among freshwater mussels (Bivalvia; Unionoidea). Journal of Animal Ecology 63: 933–944.

Bauer, G. 1998. Allocation policy of female freshwater pearl mussels. Oecologia 117: 90–94.

Bauer, G., and C. Vogel. 1987. The parasitic stage of the freshwater pearl mussel (*Margaritifera margaritifera* L.) I. Host response to glochidiosis. Archiv für Hydrobiologie Supplementband 76: 393–402.

Bauer, G., and K. Wächtler (eds.). 2001. Ecology and evolution of the freshwater mussels Unionoida. Springer-Verlag, Berlin. 394 pp.

Beasley, C. R. 2001. The impact of exploitation on freshwater mussels (Bivalvia: Hyriidae) in the Tocantins River, Brazil. Studies on Neotropical Fauna and Environment 36:159–165.

Beaty, B.B., and R.J. Neves. 2004. Use of a natural river water flow-through culture system for rearing juvenile freshwater mussels (Bivalvia: Unionidae) and evaluation of the effects of substrate size, temperature, and stocking density. American Malacological Bulletin 19: 15–23.

Beck, K., and R.J. Neves. 2003. An evaluation of selective feeding by three age-groups of the rainbow mussel *Villosa iris*. North American Journal of Aquaculture 65: 203–209.

Beck, M.B. 1996. Transient pollution events: acute risks to the aquatic environment. Water Science and Technology 33: 1–15.

Beckett, D.C., B.W. Green, S.A. Thomas, and A.C. Miller. 1996. Epizoic invertebrate communities on upper Mississippi River unionid bivalves. American Midland Naturalist 135: 102–114.

Benke, A.C. 1990. A perspective on America's vanishing streams. Journal of the North American Benthological Society 9: 77–88.

Benke, A.C., and C.E. Cushing (eds.). 2005. Rivers of North America. Elsevier, Amsterdam. 1144 pp.

Berrow, S.D. 1991. Predation by the hooded crow *Corvus corone cornix* on freshwater pearl mussels *Margaritifera margaritifera*. Irish Naturalists' Journal 23: 492–493.

Bigham, S.E. 2002. Host specificity of freshwater mussels: a critical factor in conservation. M.S. thesis, Southwest Missouri State University, Springfield, IL. 48 pages.

Bisbee, G.D. 1984. Ingestion of phytoplankton by two species of freshwater mussels, the black sandshell, *Ligumia recta*, and the three ridger (sic), *Amblema plicata*, from the Wisconsin River in Oneida County, Wisconsin. Bios 58: 219–225.

Blaise, C., F. Gagné, M. Salazar, S. Salazar, S. Trottier, and P.D. Hansen. 2003. Experimentally induced feminization of freshwater mussels after long-term exposure to municipal effluent. Fresenius Environmental Bulletin 12: 865–870.

Blalock, H.N., and J.B. Sickel. 1996. Changes in mussel (Bivalvia: Unionidae) fauna within the Kentucky portion of Lake Barkley since impoundment of the lower Cumberland River. American Malacological Bulletin 13: 111–116.

Blažek, R., and M. Gelnar. 2006. Temporal and spatial distribution of glochidial larval stages of European unionid mussels (Mollusca: Unionidae) on host fishes. Folia Parasitologica 53: 98–106.

Bogan, A.E. 1993. Freshwater bivalve extinctions (Mollusca: Unionoida): a search for causes. American Zoologist 33: 599–609.

Bohnsack, J.A. 1989. Are high densities of fishes at artificial reefs the result of habitat limitation or behavioral preference? Bulletin of Marine Science 44: 631–645.

Bolden, S.R., and K.M. Brown. 2002. Role of stream, habitat, and density in predicting translocation success in the threatened Louisiana pearlshell, *Margaritifera hembeli* (Conrad). Journal of the North American Benthological Society 21: 89–96.

Bormann, F.H. 2005. The end of reductionism? Frontiers in Ecology and the Environment 3: 472.

Bovbjerg, R.V. 1956. Mammalian predation on mussels. Proceedings of the Iowa Academy of Science 63: 737–740.

Bovbjerg, R.V. 1957. Feeding related to mussel activity. Proceedings of the Iowa Academy of Sciences 64: 650–653.

Bowen, Z.H., S.P. Malvestuto, W.D. Davies, and J.H. Crance. 1994. Evaluation of the mussel fishery in Wheeler Reservoir, Tennessee River. Journal of Freshwater Ecology 9: 313–319.

Brainwood, M., S. Burgin, and M. Byrne. 2006. Is the decline of freshwater mussel populations in a regulated coastal river in southeastern Australia linked with human modification of habitat? Aquatic Conservation: Marine and Freshwater Ecosystems 16: 501–516.

Brim Box, J., R.M. Dorazio, and W.D. Liddell. 2002. Relationships between streambed substrate characteristics and freshwater mussels (Bivalvia: Unionidae) in Coastal Plain streams. Journal of the North American Benthological Society 21: 253–260.

Brim Box, J., and J. Mossa. 1999. Sediment, land use, and freshwater mussels: Prospects and problems. Journal of the North American Benthological Society 18: 99–117.

Brönmark, C., and B. Malmqvist. 1982. Resource partitioning between unionid mussels in a Swedish lake outlet. Holarctic Ecology 5: 389–395.

Brookes, A. 1996. River channel change. Pages 221–242 *In*: G. Petts and P. Calow (eds.). River flows and channel forms. Blackwell Science, Oxford.

Brown, J.H., and M.V. Lomolino. 1998. Biogeography. Second edition. Sinauer, Sunderland, MA. 691 pp.

Brown, K.M. 1998. The role of shell strength in selective foraging by crayfish for gastropod prey. Freshwater Biology 40: 255–260.

Buddensiek, V., H. Engel, S. Fleischauer-Rössing, and K. Wächtler. 1993. Studies on the chemistry of interstitial water taken from defined horizons in the fine sediments of bivalve habitats in several northern German lowland waters. II: Microhabitats of *Margaritifera margaritifera* L., *Unio crassus* (Philipsson) and *Unio tumidus* Philipsson. Archiv für Hydrobiologie 127: 151–166.

Bunn, S.E., and P.I. Boon. 1993. What sources of organic carbon drive food webs in billabongs? A study based on stable isotope analysis. Oecologia 96: 85–94.

Burgin, A.J., and S.K. Hamilton. 2007. Have we overemphasized the role of denitrification in aquatic ecosystems? A review of nitrate removal pathways. Frontiers in Ecology and the Environment 5: 89–96.

Burla, H., H.-J. Schenker, and W. Stahel. 1974. Das Dispersionsmuster von Teichmuscheln (*Anodonta*) im Zürichsee. Oecologia 17: 131–140.

Burnham, K.P., and D.R. Anderson. 2002. Model selection and multimodel inference: A practical information-theoretic approach. Second edition. Springer-Verlag, New York. 488 pp.

Byrne, M. 1998. Reproduction of river and lake populations of *Hyridella depressa* (Unionacea: Hyriidae) in New South Wales: Implications for their conservation. Hydrobiologia 389: 29–43.

Caley, M.J., M.H. Carr, M.A. Hixon, T.P. Hughes, G.P. Jones, and B.A. Menge. 1996. Recruitment and the local dynamics of open marine populations. Annual Review of Ecology and Systematics 27: 477–500.

Call, R.E. 1900. A descriptive illustrated catalogue of the Mollusca of Indiana. Twenty-fourth Annual Report of the Indiana Department of Geology and Natural Resources: 335–535 + plates 1–78.

Campbell, D.C., J.M. Serb, J.E. Buhay, K.J. Roe, R.L. Minton, and C. Lydeard. 2005. Phylogeny of North American amblemines: Prodigious polyphyly proves pervasive across genera. Invertebrate Biology 124: 131–164.

Caraco, N.F., J.J. Cole, S.E.G. Findlay, D.T. Fischer, G.G. Lampman, M.L. Pace and D.L. Strayer. 2000. Dissolved oxygen declines in the Hudson River associated with the invasion of the zebra mussel (*Dreissena polymorpha*). Environmental Science and Technology 34: 1204–1210.

Caraco, N.F., J.J. Cole, P.A. Raymond, D.L. Strayer, M.L. Pace, S.E.G. Findlay, and D.T. Fischer. 1997. Zebra mussel invasion in a large, turbid river: phytoplankton response to increased grazing. Ecology 78: 588–602.

Caraco, N.F., J.J. Cole, and D.L. Strayer. 2006. Top down control from the bottom: Regulation of eutrophication in a large river by benthic grazing. Limnology and Oceanography 51: 664–670.

Carpenter, S.R. (ed.). 1988. Complex interactions in lake communities. Springer-Verlag, New York. 283 pp.

Carpenter, S.R., N.F. Caraco, D.L. Correll, R.W. Howarth, A.N. Sharpley, and V.H. Smith. 1998. Nonpoint pollution of surface waters with phosphorus and nitrogen. Ecological Applications 8: 559–568.

Carpenter, S.R., D.L. Christensen, J.J. Cole, K.L. Cottingham, X. He, J.R. Hodgson, J.F. Kitchell, S.E. Knight, M.L. Pace, D.M. Post, D.E. Schindler, and N. Voichick. 1995. Biological control of eutrophication in lakes. Environmental Science and Technology 29: 784–786.

Carpenter, S.R., and J.F. Kitchell (eds.). 1996. The trophic cascade in lakes. Cambridge University Press, New York. 399 pp.

Carpenter, S.R., J.F. Kitchell, and J.R. Hodgson. 1985. Cascading trophic interactions and lake productivity. BioScience 35: 634–639.

Chamberlain, T.K. 1931. Annual growth of fresh-water mussels. Bulletin of the United States Bureau of Fisheries 46: 713–738.

Chambers, P.A., E.E. Prepas, and K. Gibson. 1992. Temporal and spatial dynamics in riverbed chemistry: The influence of flow and sediment composition. Canadian Journal of Fisheries and Aquatic Sciences 49: 2128–2140.

Chatelain, R., and J. Chabot. 1983. Utilisation d'accumulation de coquilles d'Unionidae comme frayeres par le touladi. Naturaliste Canadiene 110: 363–365.

Christian, A.D., B.N. Smith, D.J. Berg, J.C. Smoot, and R.H. Findlay. 2004. Trophic position and potential food sources of 2 species of unionid bivalves (Mollusca: Unionidae) in 2 small Ohio streams. Journal of the North American Benthological Society 23: 101–113.

Churchill, E.P., and S.I. Lewis. 1924. Food and feeding in fresh-water mussels. Bulletin of the Bureau of Fisheries 39: 439–471.

Claassen, C. 1994. Washboards, pigtoes, and muckets: historic musseling in the Mississippi watershed. Historical Archaeology 28: 1–145.

Clarke, A.H. 1973. The freshwater molluscs of the Canadian Interior Basin. Malacologia 13: 1–509.

Clarke, A.H. 1981. The freshwater molluscs of Canada. National Museums of Canada, Ottawa. 446 pp.

Clayton, J.L., C.W. Stihler, and J.L. Wallace. 2001. Status of and potential impacts to the freshwater bivalves (Unionidae) in Patterson Creek, West Virginia. Northeastern Naturalist 8: 179–188.

Cloern, J.E. 2001. Our evolving conceptual model of the coastal eutrophication problem. Marine Ecology Progress Series 210: 223–253.

Coker, R.E., A.F. Shira, H.W. Clark, and A.D. Howard. 1921. Natural history and propagation of fresh-water mussels. Bulletin of the Bureau of Fisheries 37: 77–181.

Conners, D.E., and M.C. Black. 2004. Evaluation of lethality and genotoxicity in the freshwater mussel *Utterbackia imbecillis* (Bivalvia: Unionidae) exposed singly and in combination to chemicals used in lawn care. Archives of Environmental Contamination and Toxicology 46: 362–371.

Convey, L.E., J.M. Hanson, and W.C. Mackay. 1989. Size-selective predation on unionid clams by muskrats. Journal of Wildlife Management 53: 654–657.

Corey, C.A. 2003. Freshwater mussel behavior: displays by members of the genus *Ligumia* and habitat selection by juveniles of two species. M.S. Thesis, State University of New York, Albany, NY. 59 pp. + 1 CD.

Corey, C.A., R. Dowling, and D.L. Strayer. 2006. Display behavior of *Ligumia* (Bivalvia: Unionidae). Northeastern Naturalist 13: 319–332.

Crosby, N.D., and R.G.B. Reid. 1971. Relationships between food, phylogeny, and cellulose digestion in Bivalvia. Canadian Journal of Zoology 49: 617–622.

Crossman, J.S., and J. Cairns. 1973. Aquatic invertebrate recovery in the Clinch River following hazardous spills and floods. Research Bulletin of the Virginia Water Resources Research Center 63, Blacksburg, VA. 66 pp.

Cummings, K.S., and A.E. Bogan. 2006. Unionoida: freshwater mussels. Pages 313–325 *In*: C.F. Sturm, T.A. Pearce, and A. Valdés (eds.). The mollusks: A guide to their study, collection, and preservation. American Malacological Society, Pittsburgh, PA. 445 pp.

Cummings, K.S., and C.A. Mayer. 1992. Field guide to freshwater mussels of the Midwest. Illinois Natural History Survey Manual 5, Champaign, IL. 194 pp.

Cummings, K.S., and G.T. Watters. 2005. Mussel/host database. http://128.146.250.63/Musselhost/. Accessed 14 November 2005.

Cunjak, R.A., and S.E. McGladdery. 1991. The parasite-host relationship of glochidia (Mollusca: Margaritiferidae) on the gills of young-of-the-year Atlantic salmon (*Salmo salar*). Canadian Journal of Zoology 69: 353–359.

Cvancara, A.M. 1972. Lake mussel distribution as determined with SCUBA. Ecology 53: 154–157.

Dame, R.F. 1996. Ecology of marine bivalves: An ecosystem approach. CRC Press, Boca Raton, FL. 254 pp.

Daniels, R.A. 2001. Untested assumptions: The role of canals in the dispersal of sea lamprey, alewife, and other fishes in the eastern United States. Environmental Biology of Fishes 60: 309–329.

Dartnall, H.J.G., and M. Walkey. 1979. The distribution of glochidia of the Swan mussel, *Anodonta cygnea* (Mollusca) on the Three–spined stickleback *Gasterosteus aculeatus* (Pisces). Journal of Zoology 189: 31–37.

Davis, G.M. 1983. Relative roles of molecular genetics, anatomy, morphometrics and ecology in assessing relationships among North America Unionidae (Bivalvia). Pages 193–222 *In*: G.S. Oxford and D. Rollison (eds.). Protein polymorphism: Adaptive and taxonomic significance. Academic Press, London.

Davis, G.M. 1984. Genetic relationships among some North American Unionidae (Bivalvia): Sibling species, convergence, and cladistic relationships. Malacologia 25: 629–648.

Delp, A.M. 2002. Flatworm predation on juvenile freshwater mussels. M.S. thesis, Southwest Missouri State University, Springfield, MO. 31 pp.

DeMelo, R., R. France, and D.J. McQueen. 1992. Biomanipulation: hit or myth? Limnology and Oceanography 37: 192–207.

Diggins, T.P., and K.M. Stewart. 2000. Evidence of large change in unionid mussel abundance from selective muskrat predation, as inferred by shell remains left on shore. International Review of Hydrobiology 85: 505–520.

Dillon, P.J., and F.H. Rigler. 1974. The phosphorus-chlorophyll relationship in lakes. Limnology and Oceanography 19: 767–773.

Dillon, R.T. 2000. Ecology of freshwater molluscs. Cambridge University Press. 509 pp.

Di Maio, J., and L.D. Corkum. 1995. Relationship between the spatial distribution of fresh-water mussels (Bivalvia, Unionidae) and the hydrological variability of rivers. Canadian Journal of Zoology 73: 663–671.

Dimock, R.V. 2000. Oxygen consumption by juvenile *Pyganodon cataracta* (Bivalvia: Unionidae) in response to declining oxygen tension. Pages 1–8 *in*: R.A. Tankersley, D.I. Warmolts, G.T. Watters, B.J. Armitage, P.D. Johnson, and R.S. Butler (eds.). Freshwater Mollusk Symposia Proceedings, Ohio Biological Survey Special Publication, Columbus, OH.

Dimock, R.V., and A.H. Wright. 1993. Sensitivity of juvenile freshwater mussels to hypoxic, thermal and acid stress. Journal of the Elisha Mitchell Scientific Society 109: 183–192.

Dodd, B.J., M.C. Barnhart, C.L. Rogers-Lowery, T.B. Fobian, and R.V. Dimock. 2005. Cross- resistance of largemouth bass to glochidia of unionid mussels. Journal of Parasitology 91: 1064–1072.

Dodd, B.J., M.C. Barnhart, C.L. Rogers-Lowery, T.B. Fobian, and R.V. Dimock. 2006. The persistence of acquired immunity of largemouth bass *Micropterus salmoides* to glochidia larvae of *Lampsilis reeveiana* (Bivalvia, Unionidae). Fish and Shellfish Immunology 21: 473–484.

Dodge, K.E. 1998. River Raisin assessment. Michigan Department of Natural Resources, Fisheries Division Special Report 23. Available at http://www.michigandnr.com/PUBLICATIONS/PDFS/ifr/ifrlibra/special/reports/sr23/sr23Text.pdf.

Downing, J.A., and W.L. Downing. 1992. Spatial aggregation, precision, and power in surveys of freshwater mussel populations. Canadian Journal of Fisheries and Aquatic Sciences 49: 985–991.

Downing, J.A., and C. Plante. 1993. Production of fish populations in lakes. Canadian Journal of Fisheries and Aquatic Sciences 50: 110–120.

Downing, J.A., Y. Rochon, M. Pérusse, and H. Harvey. 1993. Spatial aggregation, body size, and reproductive success in the freshwater mussel *Elliptio complanata*. Journal of the North American Benthological Society 12: 148–156.

Doyle, M.W., E.H. Stanley, and J.M. Harbor. 2003. Channel adjustments following two dam removals in Wisconsin. Water Resources Research 39: Article Number 1011.

Drake, J.M., and J.M. Bossenbroek. 2004. The potential distribution of zebra mussels in the United States. BioScience 54: 931–941.

Dudgeon, D., A.H. Arthington, M.O. Gessner, Z.-I. Kawabata, D.J. Knowler, C. Lévêque, R.J. Naiman, A.-H. Prieur-Richard, D. Soto, M.L.J. Stiassny, and C.A. Sullivan. 2005. Freshwater biodiversity: Importance, threats, status and conservation challenges. Biological Reviews 81: 163–182.

Dudgeon, D., and B. Morton. 1984. Site selection and attachment duration of *Anodonta woodiana* (Bivalvia: Unionacea) glochidia on fish hosts. Journal of Zoology 204: 355–362.

Duobinis-Gray, L., E. Urban, J. Sickel, D. Owen, and W. Maddox. 1991. Aspidogastrid (Trematoda) parasites on unionid (Bivalvia) molluscs in Kentucky Lake. Journal of the Helminthological Society of Washington 58: 167–170. (Original not seen; cited by Dillon 2000).

Edwards, D.D., and R.V. Dimock. 1988. A comparison of the population dynamics of *Unionicola formosa* from two anodontine bivalves in a North Carolina farm pond. Journal of the Elisha Mitchell Scientific Society 104: 90–98.

Edwards, D.D., and M.F. Vidrine. 2006. Host specificity among *Unionicola* spp. (Acari: Unionicolidae) parasitizing freshwater mussels. Journal of Parasitology 92: 977–983.

Edwards, W.J., C.R. Rehmann, E. McDonald, and D.A. Culver. 2005. The impact of a benthic filter-feeder: Limitations imposed by physical transport of algae to the benthos. Canadian Journal of Fisheries and Aquatic Sciences 62: 205–214.

Effler, S.W., S.R. Boone, C. Siegfried, and S.L. Ashby. 1998. Dynamics of zebra mussel oxygen demand in Seneca River, New York. Environmental Science and Technology 32: 807–812.

Effler, S.W., and C. Siegfried. 1994. Zebra mussel (*Dreissena polymorpha*) populations in the Seneca River, New York: Impact on oxygen resources. Environmental Science and Technology 28: 2216–2221.

Efron, B., and R.J. Tibshirani. 1993. An introduction to the bootstrap. Chapman and Hall, New York. 436 pp.

Elderkin, C.L., A.D. Christian, C.C. Vaughn, J.L. Metcalfe-Smith, and D.J. Berg. 2007. Population genetics of the freshwater mussel, *Amblema plicata* (Say 1817) (Bivalvia: Unionidae): Evidence of high dispersal and post-glacial colonization. Conservation Genetics 8: 355–372.

Emerson, K., R.C. Russo, R.E. Lund, and R.V. Thurston. 1975. Aqueous ammonia equilibrium calculations: Effect of pH and temperature. Journal of the Fisheries Research Board of Canada 32: 2379–2383.

Englund, G. 1999. Effects of fish on the local abundance of crayfish in stream pools. Oikos 87: 48–56.

EPA. 2006. U.S. Environmental Protection Agency STORET. http://www.epa.gov/storet/. Accessed 26 July 2006.

Fagan, W.F., P.J. Unmack, C. Burgess, and W.L. Minckley. 2002. Rarity, fragmentation, and extinction risk in desert fishes. Ecology 83: 3250–3256.

Findlay, S., M. L. Pace, and D. T. Fischer. 1998. Response of heterotrophic planktonic bacteria to the zebra mussel invasion of the tidal freshwater Hudson River. Microbial Ecology 36: 131–140.

Finlay, J.C., S. Khandwala, and M.E. Power. 2002. Spatial scales of carbon flow in a river food web. Ecology 83: 1845–1859.

Fisher, J.B., and M.J.S. Tevesz. 1976. Distribution and population density of *Elliptio complanata* (Mollusca) in Lake Pocotopaug, Connecticut. The Veliger 18: 332–338.

Flack, V.F., and P.C. Chang. 1987. Frequency of selecting noise variable in subset regression analysis: a simulation study. The American Statistician 41: 84–86.

Fortin, M.-J., and M. Dale. 2005. Spatial analysis: A guide for ecologists. Cambridge University Press, Cambridge, UK. 365 pp.

Fraley, S.J., and S.A. Ahlstedt. 2000. The recent decline of the native mussels (Unionidae) of Copper Creek, Russell and Scott Counties, Virginia. Pages 189–195 In: Tankersley, R.A., D.I. Warmolts, G.T. Watters, B.J. Armitage, P.D. Johnson, and R.S. Butler (eds.). Freshwater Mollusk Symposia Proceedings. Ohio Biological Survey, Columbus, OH.

Fukuhara, S., and Y. Nagata. 1995. Estimation of the factors determining the intervals among individuals of the freshwater mussel *Anodonta woodiana* Lea (Bivalvia: Unionidae) in a small pond. Venus 54: 317–327 (In Japanese with an English summary).

Fuller, P.L., L.G. Nico, and J.D. Williams. 1999. Nonindigenous fishes introduced into inland waters of the United States. American Fisheries Society Special Publication 27: 1–613.

Fuller, S.L.H. 1974. Clams and mussels (Mollusca: Bivalvia). Pages 215–273 In: C.W. Hart and S.L.H. Fuller (eds.). Pollution ecology of freshwater invertebrates. Academic Press, New York.

Gagné, F., M. Fournier, and C. Blaise. 2004. Serotonergic effects of municipal effluents: Induced spawning activity in freshwater mussels. Fresenius Environmental Bulletin 13: 1099–1103.

Gagnon, P.M., S.W. Golladay, W.K. Michener, and M.C. Freeman. 2004. Drought responses of freshwater mussels (Unionidae) in coastal plain tributaries of the Flint River basin, Georgia. Journal of Freshwater Ecology 19: 667–679.

Gagnon, P., W. Michener, M. Freeman, and J. Brim Box. 2006. Unionid habitat and assemblage composition in Coastal Plain tributaries of Flint River (Georgia). Southeastern Naturalist 5: 31–52.

Gangloff, M.M., and J.W. Feminella. 2007. Stream bed geomorphology influences mussel abundance in southern Appalachian streams, U.S.A. Freshwater Biology 52: 64–74.

Garner, J.T., and S.W. McGregor. 2001. Current status of freshwater mussels (Unionidae, Margaritiferidae) in the Muscle Shoals area of Tennessee River in Alabama (Muscle Shoals revisited again). American Malacological Bulletin 16: 155–170.

Gaston, K.J. 2003. The structure and dynamics of geographic ranges. Oxford University Press, Oxford, UK. 266 pp.

Gatenby, C.M., R.J. Neves, and B.C. Parker. 1996. Influence of sediment and algal food on cultured juvenile freshwater mussels. Journal of the North American Benthological Society 15: 597–609.

Gatenby, C.M., B.C. Parker, and R.J. Neves. 1997. Growth and survival of juvenile rainbow mussels, *Villosa iris* (Lea, 1829) (Bivalvia: Unionidae), reared on algal diets and sediment. American Malacological Bulletin 14: 57–66.

Geist, J., and K. Auerswald. 2007. Physicochemical streambed characteristics and recruitment of the freshwater pearl mussel (*Margaritifera margaritifera*). Freshwater Biology 52: 2299–2316.

Geist, J., and R. Kuehn. 2005. Genetic diversity and differentiation of central European freshwater pearl mussel (*Margaritifera margaritifera* L.) populations: Implications for conservation and management. Molecular Ecology 14: 425–439.

Geist, J., M. Porkka, and R. Kuehn. 2006. The status of host fish populations and fish species richness in European freshwater pearl mussel (*Margaritifera margaritifera*) streams. Aquatic Conservation: Marine and Freshwater Ecosystems 16: 251–266.

Ghent, A.W., R. Singer, and L. Johnson-Singer. 1978. Depth distributions determined with SCUBA, and associated studies of the freshwater unionid clams *Elliptio complanata* and *Anodonta grandis* in Lake Bernard, Ontario. Canadian Journal of Zoology 56: 1654–1663.

Gittings, T., D. O'Keefe, F. Gallagher, J. Finn, and T. O'Mahony. 1998. Longitudinal variation in abundance of a freshwater pearl mussel *Margaritifera margaritifera* population in relation to riverine habitats. Biology and Environment: Proceedings of the Royal Irish Academy 98B: 171–178.

Golladay, S.W., P. Gagnon, M. Kearns, J.M. Battle, and D.W. Hicks. 2004. Response of freshwater mussel assemblages (Bivalvia: Unionidae) to a record drought in the Gulf Coastal Plain of southwestern Georgia. Journal of the North American Benthological Society 23: 494–506.

Gordon, M.E., and J.B. Layzer. 1993. Glochidial host of *Alasmidonta atropurpurea* (Bivalvia: Unionoidea, Unionidae). Transactions of the American Microscopical Society 112: 145–150.

Gordon, N.D., T.A. McMahon, and B.L. Finlayson. 1992. Stream hydrology: an introduction for ecologists. John Wiley, Chichester. 526 pp.

Goudreau, S.E., R.J. Neves, and R.J. Sheehan. 1993. Effects of wastewater treatment plant effluents on freshwater mollusks in the upper Clinch River, Virginia, USA. Hydrobiologia 252: 211–230.

Graf, D.L. 2000. The Etherioidea revisited: A phylogenetic analysis of hyriid relationships (Mollusca: Bivalvia: Paleoheterodonta: Unionoida). Occasional Papers of the University of Michigan Museum of Zoology 729: 1–21.

Graf, D.L., and K.S. Cummings. 2006. Palaeoheterodont diversity (Mollusca: Trigonioida + Unionoida): What we know and what we wish we knew about freshwater mussel evolution. Zoological Journal of the Linnean Society 148: 343–394.

Graf, D.L., and D. Ó Foighil. 2000. Molecular phylogenetic analysis of 28S rRNA supports a Gondwanan origin for Australasian Hyriidae (Mollusca: Bivalvia: Unionoida). Vie et Milieu 50: 245–254.

Green, R.H. 1980. Role of a unionid clam population in the calcium budget of a small Arctic lake. Canadian Journal of Fisheries and Aquatic Sciences 37: 219–224.

Greene, H.W. 2005. Organisms in nature as a central focus for biology. Trends in Ecology and Evolution 20: 23–27.

Grossman, G.D., J. Hill, and J.T. Petty. 1995. Observations on habitat structure, population regulation, and habitat use in evolutionarily significant units: a landscape perspective for lotic systems. American Fisheries Society Symposium 17: 381–391.

Grossman, G.D., G.P. Jones, and W.J. Seaman. 1997. Do artificial reefs increase regional fish production? A review of existing data. Fisheries 22(4): 17–23.

Guisan, A., and W. Thuiller. 2005. Predicting species distributions: Offering more than simple habitat models. Ecology Letters 8: 993–1009.

Gutierrez, J.L., C.G. Jones, D.L. Strayer, and O.O. Iribarne. 2003. Mollusks as ecosystem engineers: Their functional roles as shell producers in aquatic habitats. Oikos 101: 79–90.

Guttinger, H., and W. Stumm. 1992. Ecotoxicology: An analysis of the Rhine pollution caused by the Sandoz chemical accident, 1986. Interdisciplinary Science Reviews 17: 127–136.

Haag, W.R., D.J. Berg, D.W. Garton, and J.L. Farris. 1993. Reduced survival and fitness in native bivalves in response to fouling by the introduced zebra mussel (*Dreissena polymorpha*) in western Lake Erie. Canadian Journal of Fisheries and Aquatic Sciences 50: 13–19.

Haag, W.R., R.S. Butler, and P.W. Hartfield. 1995. An extraordinary reproductive strategy in freshwater bivalves: Prey mimicry to facilitate larval dispersal. Freshwater Biology 43: 471–476.

Haag, W.R., and M.L. Warren. 1997. Host fishes and reproductive biology of six freshwater mussel species from the Mobile Basin, USA. Journal of the North American Benthological Society 16: 576–585.

Haag, W.R., and M.L. Warren. 1998. Role of ecological factors and reproductive strategies in structuring freshwater mussel communities. Canadian Journal of Fisheries and Aquatic Sciences 55: 297–306.

Haag, W.R., and M.L. Warren. 2003. Host fishes and infection strategies of freshwater mussels in large Mobile basin streams, USA. Journal of the North American Benthological Society 22: 78–91.

Haag, W.R., M.L. Warren, and M. Shillingsford. 1999. Host fishes and host-attracting behavior of *Lampsilis altilis* and *Villosa vibex* (Bivalvia: Unionidae). American Midland Naturalist 141: 149–157.

Haag, W.R., and J. L Staton. 2003. Variation in fecundity and other reproductive traits in freshwater mussels. Freshwater Biology 48: 2118–2130.

Hakanson, L. 1995. Optimal size of predictive models. Ecological Modelling 78: 195–204.

Hanski, I., A. Moilanen, and M. Gyllenberg. 1996. Minimum viable population size. American Naturalist 147: 527–541.

Hanson, J.M., W.C. Mackay, and E.E. Prepas. 1988. Population size, growth, and production of a unionid clam, *Anodonta grandis simpsoniana*, in a small, deep boreal forest lake in central Alberta. Canadian Journal of Zoology 66: 247–253.

Hanson, J.M., W.C. Mackay, and E.E. Prepas. 1989. Effect of size-selective predation by muskrats (*Ondatra zebithicus*) on a population of unionid clams (*Anodonta grandis simpsoniana*). Journal of Animal Ecology 58: 15–28.

Hardison, B.S., and J.B. Layzer. 2001. Relations between complex hydraulics and the localized distribution of mussels in three regulated rivers. Regulated Rivers: Research and Management 17: 77–84.

Hart, R.A., and J.W. Grier. 2004. Simulation models of harvested and zebra mussel-colonized threeridge mussel populations in the upper Mississippi River. American Midland Naturalist 151: 301–317.

Hart, R.A., J.W. Grier, A.C. Miller, and M. Davis. 2001. Empirically derived survival rates of a native mussel, *Amblema plicata*, in the Mississippi and Otter Tail Rivers, Minnesota. American Midland Naturalist 146: 254–263.

Hartfield, P. 1993. Headcuts and their effect on freshwater mussels. Pages 131–141 *In*: K.S. Cummings, A.C. Buchanan, and L.M. Koch (eds.). Conservation and management of freshwater mussels. Upper Mississippi River Conservation Committee, Rock Island, IL.

Hartfield, P., and E. Hartfield. 1996. Observations on the conglutinates of *Ptychobranchus greeni* (Conrad, 1834) (Mollusca: Bivalvia: Unionoidea). American Midland Naturalist 135: 370–375.

Hastie, L.C., P.J. Boon, M.R. Young, and S. Way. 2001. The effects of a major flood on an endangered freshwater mussel population. Biological Conservation 98: 107–115.

Hastie, L.C., P.J. Cosgrove, N. Ellis, and M.J. Gaywood. 2003. The threat of climate change to freshwater pearl mussel populations. Ambio 32: 40–46.

Hastie, L.C., and M.R. Young. 2001. Freshwater pearl mussel (*Margaritifera margaritifera*) glochidiosis in wild and farmed salmonid stocks in Scotland. Hydrobiologia 445: 109–119.

Hastie, L.C., and M.R. Young. 2003. Timing of spawning and glochidial release in Scottish freshwater pearl mussel (*Margaritifera margaritifera*) populations. Freshwater Biology 48: 2107–2117.

Hastie, L.C., M.R. Young, P.J. Boon, P.J. Cosgrove, and B. Henninger. 2000. Sizes, densities, and age structures of Scottish *Margaritifera margaritifera* (L.) populations. Aquatic Conservation: Marine and Freshwater Ecosystems 10: 229–247.

Haukioja, E., and T. Hakala. 1974. Vertical distribution of freshwater mussels (Pelecypoda, Unionidae) in southwestern Finland. Annales Zoologici Fennici 11: 127–130.

Headlee, T.J. 1906. Ecological notes on the mussels of Winona, Pike, and Center lakes of Kosciusko County, Indiana. Biological Bulletin 11: 305–319.

Heathwaite, A.L., P.J. Johnes, and N.E. Peters. 1996. Trends in nutrients. Hydrological Processes 10: 263–293.

Heinricher, J.R., and J.B. Layzer. 1999. Reproduction by individuals of a non-reproducing population of *Megalonaias nervosa* (Mollusca: Unionidae) following translocation. American Midland Naturalist 141: 140–148.

Henley, W.F., and R.J. Neves. 1999. Recovery status of freshwater mussels (Bivalvia: Unionidae) in the North Fork Holston River, Virginia. American Malacological Bulletin 15: 65–73.

Hilborn, R., and M. Mangel. 1997. The ecological detective: Confronting models with data. Princeton University Press, Princeton, NJ. 330 pp.

Hobbs, H.H. III, J.P. Jass, and J.V. Huner. 1989. A review of global crayfish introductions, with particular emphasis on two North American species (Decapoda, Cambaridae). Crustaceana 56: 299–316.

Hochwald, S. 2001. Plasticity of life-history traits in *Unio crassus*. Pages 127–141 In: G. Bauer and K. Wächtler (eds.). Ecology and evolution of the freshwater mussels Unionoida. Springer-Verlag, Berlin.

Hoeh, W.R., A.E. Bogan, and W.H. Heard. 2001. A phylogenetic perspective on the evolution of morphological and reproductive characters in the Unionoida. Pages 257–280 In: G. Bauer and K. Wächtler (eds.). Ecology and evolution of the freshwater mussels Unionoida. Springer-Verlag, Berlin.

Holland-Bartels, L.E. 1990. Physical factors and their influence on the mussel fauna of a main channel border habitat of the upper Mississippi River. Journal of the North American Benthological Society 9: 327–335.

Holt, R.D., and J.H. Lawton. 1994. The ecological consequences of shared natural enemies. Annual Review of Ecology and Systematics 25: 495–520.

Hove, M., M. Berg, K. Dietrich, C. Gonzalez, D. Hornbach, K. Juleen, M. Ledford, M. Marzec, M. McGill, C. Nelson, B.J. Ritger, J. Selander, and A. Kapuscinski. 2003. High school students participate in snuffbox host suitability trials. Ellipsaria 5: 19–20.

Hove, M.C., K.R. Hillegass, J.E. Kurth, V.E. Pepi, C.J. Lee, K.A. Knudsen, A.R. Kapuscinski, P.A. Kahoney, and M.V. Bomier. 2000. Considerations for conducting host suitability studies. Pages 27–34 *In*: R.A. Tankersley, D.I. Warmolts, G.T. Watters, B.J. Armitage, P.D. Johnson, and R.S. Butler (eds.). Freshwater mollusk symposium proceedings. Ohio Biological Survey, Columbus, OH.

Howard, A.D. 1914. Experiments in propagation of fresh-water mussels of the *Quadrula* group. Bureau of Fisheries Document 801: 1–52 + 6 plates. Washington, DC.

Howard, A.D. 1951. A river mussel parasitic on a salamander. Natural History Miscellanea 77: 1–6.

Howard, J.K., and K.M. Cuffey. 2003. Freshwater mussels in a California North Coast Range river: Occurrence, distribution, and controls. Journal of the North American Benthological Society 22: 63–77.

Howard, J.K., and K.M. Cuffey. 2006. Factors controlling the age structure of *Margaritifera falcata* in 2 California northern streams. Journal of the North American Benthological Society 25: 677–690.

Howells, R.G., R.W. Neck, and H.D. Murray. 1996. Freshwater mussels of Texas. Texas Parks and Wildlife Department, Austin, TX. 218 pp.

Hruška, J. 1992. The freshwater pearl mussel in South Bohemia: Evaluation of the effect of temperature on reproduction, growth and age structure of the population. Archiv für Hydrobiologie 126: 181–191.

Huebner, J.D. 1982. Seasonal variation in two species of unionid clams from Manitoba, Canada: Respiration. Canadian Journal of Zoology 60: 560–564.

Huebner, J.D., D.F. Malloy, and K. Donkersloot. 1990. Population ecology of the freshwater mussel *Anodonta grandis grandis* in a Precambrian Shield lake. Canadian Journal of Zoology 68: 1931–1941.

Huehner, M.K. 1984. Aspidogastrid trematodes from freshwater mussels in Missouri with notes on the life history of *Cotylaspis insignis*. Proceedings of the Helminthological Society of Washington 51: 270–274. (original not seen; cited by Dillon 2000).

Huehner, M.K. 1996. A survey of unionid mussels of the Grand River's Wild and Scenic section. Final Report to the Ohio Department of Natural Resources. 7 pp. + 3 tables + 3 maps.

Huehner, M.K. 1997. A survey of unionid mussels of the Grand River's Scenic section. Final Report to the Ohio Department of Natural Resources. 6 pp. + 3 tables + 3 appendices.

Huehner, M.K. 1999. Mussel survey of Grand River headwaters and distribution of the clubshell in Pymatuning Creek. Final Report to the Ohio Department of Natural Resources. 7 pp. + 4 tables + 11 figures.

Huff, S.W., D. Campbell, D.L. Gustafson, C. Lydeard, C.R. Altaba, and G. Giribet. 2004. Investigations into the phylogenetic relationships of freshwater pearl mussels (Bivalvia: Margaritiferidae) based on molecular data: Implications for their taxonomy and biogeography. Journal of Molluscan Studies 70: 379–388.

Hughes, J.M. 2007. Constraints on recovery: Using molecular methods to study connectivity of aquatic biota in rivers and streams. Freshwater Biology 52: 616–631.

Hughes, J., A.M. Baker, C. Bartlett, S. Bunn, K. Goudkamp, and J. Somerville. 2004. Past and present patterns of connectivity among populations of four cryptic species of freshwater mussels *Velesunio* spp. (Hyriidae) in central Australia. Molecular Ecology 13: 3197–3212.

Hughes, M.H., and P.W. Parmalee. 1999. Prehistoric and modern freshwater mussel (Mollusca: Bivalvia: Unionoidea) faunas of the Tennessee River: Alabama, Kentucky, and Tennessee. Regulated Rivers: Research and Management 15: 25–42.

Hynes, H.B.N. 1960. The biology of polluted waters. Liverpool University Press. 202 pp.

IUCN. 2006. 2006 IUCN Red List of threatened species. www.iucnredlist.org. Accessed 30 August 2006.

Jackson, C.R., J.K. Martin, D.S. Leigh, and L.T. West. 2005. A southeastern piedmont watershed sediment budget: Evidence for a multi-millennial agricultural legacy. Journal of Soil and Water Conservation 60: 298–310.

Jackson, R. B., S. R. Carpenter, C. N. Dahm, D. M. McKnight, R. J. Naiman, S. L. Postel, and S. W. Running. 2001. Water in a changing world. Ecological Applications 11:1027–1045.

Jacobson, P.J., R.J. Neves, D.S. Cherry, and J.L. Farris. 1997. Sensitivity of glochidial stages of freshwater mussels (Bivalvia: Unionidae) to copper. Environmental Toxicology and Chemistry 16: 2384–2392.

James, M.R. 1985. Distribution, biomass and production of the freshwater mussel, *Hyridella menziesi* (Gray) in Lake Taupo, New Zealand. Freshwater Biology 15: 307–314.

James, M.R. 1987. Ecology of the freshwater mussel *Hyridella menziesi* (Gray) in a small oligotrophic lake. Archiv für Hydrobiologie 108: 337–348.

Jansen, W.A. 1991. Seasonal prevalence, intensity of infestation, and distribution of glochidia of *Anodonta grandis simpsoniana* Lea on yellow perch, *Perca flavescens*. Canadian Journal of Zoology 69: 964–972.

Jansen, W., G. Bauer, and E. Zahner-Meike. 2001. Glochidial mortality in freshwater mussels. Pages 185–211 *In*: G. Bauer and K. Wächtler (eds.). Ecology and evolution of the freshwater mussels Unionoida. Springer-Verlag, Berlin.

Jansen, W.A., and J.M. Hanson. 1991. Estimates of the number of glochidia produced by clams (*Anodonta grandis simpsoniana* Lea), attaching to yellow perch (*Perca flavescens*), and surviving to various ages in Narrow Lake, Alberta. Canadian Journal of Zoology 69: 973–977.

Jeschke, J., and D.L. Strayer. 2005. Invasion success of vertebrates in Europe and North America. Proceedings of the National Academy of Sciences 102: 7198–7202.

Jirka, K.J., and R.J. Neves. 1992. Reproductive biology of 4 species of freshwater mussels (Mollusca, Unionidae) in the New River, Virginia and West Virginia. Journal of Freshwater Ecology 7: 35–44.

Johnson, P.D., and K.M. Brown. 1998. Intraspecific life history variation in the threatened Louisiana pearlshell mussel, *Margaritifera hembeli*. Freshwater Biology 40: 317–329.

Johnson, P.D., and K.M. Brown. 2000. The importance of microhabitat factors and habitat stability to the threatened Louisiana pearl shell, *Margaritifera hembeli* (Conrad). Canadian Journal of Zoology 78: 271–277.

Johnson, R.L., F.Q. Liang, C.D. Milam, and J.L. Farris. 1998. Genetic diversity and cellulolytic activity among several species of unionid bivalves in Arkansas. Journal of Shellfish Research 17: 1375–1382.

Jokela, J., and P. Mutikainen. 1995. Effect of size-dependent muskrat (*Ondatra zebithicus*) predation on the spatial distribution of a fresh-water clam, *Anodonta piscinalis* Nilsson (Unionidae, Bivalvia). Canadian Journal of Zoology 73: 1085–1094.

Jokela, J., L. Uotila, and J. Taskinen. 1993. Effect of the castrating trematode parasite *Rhipidocotyle fennica* on energy allocation of freshwater clam *Anodonta piscinalis*. Functional Ecology 7: 332–338.

Jokela, J., E.T. Valtonen, and M. Lappalainen. 1991. Development of glochidia of *Anodonta piscinalis* and their infection of fish in a small lake in northern Finland. Archiv für Hydrobiologie 120: 345–355.

Jones, J.W., R.J. Neves, S.A. Ahlstedt, and E.M. Hallerman. 2006. A holistic approach to taxonomic evaluation of two closely related endangered freshwater mussel species, the oyster mussel *Epioblasma capsaeformis* and tan riffleshell *Epioblasma florentina walkeri* (Bivalvia: Unionidae). Journal of Molluscan Studies 72: 267–283.

Karatayev, A.Y., L.E. Burlakova, and D.K. Padilla. 2006. Growth rate and longevity of *Dreissena polymorpha* (Pallas): A review and recommendations for future study. Journal of Shellfish Research 25: 23–32.

Kat, P.W. 1983. Sexual selection and simultaneous hermaphroditism among the Unionidae (Bivalvia, Mollusca). Journal of Zoology 201: 395–416.

Kelner, D.E., and B.E. Sietman. 2000. Relic populations of the ebony shell, *Fusconaia ebena* (Bivalvia: Unionidae) in the upper Mississippi River drainage. Journal of Freshwater Ecology 15: 371–377.

Kelso, B.H.L., R.V. Smith, R.J. Laughlin, and S.D. Lennox. 1997. Dissimilatory nitrate reduction in anaerobic sediments leading to river nitrite accumulation. Applied and Environmental Microbiology 63: 4679–4685.

Kesler, D.H., and R.C. Bailey. 1993. Density and ecomorphology of a freshwater mussel, *Elliptio complanata* (Bivalvia: Unionidae), in a Rhode Island lake. Journal of the North American Benthological Society 12: 259–264.

Kesler, D.H., T.J. Newton, and L. Green. 2007. Long-term monitoring of growth in the Eastern Elliptio, *Elliptio complanata* (Bivalvia: Unionidae), in Rhode Island: A transplant experiment. Journal of the North American Benthological Society 26: 123–133.

Kidd, K.A., P.J. Blanchfield, K.H. Mills, V.P. Palace, R.E. Evans, J.M. Lazorchak, and R.W. Flick. 2007. Collapse of a fish population after exposure to a synthetic estrogen. Proceedings of the National Academy of Sciences 104: 8897–8901.

King, T.L., M.S. Eackles, B. Gjetvaj, and W.R. Hoeh. 1999. Intraspecific phylogeography of *Lasmigona subviridis* (Bivalvia: Unionidae): Conservation implications of range discontinuity. Molecular Ecology 8 (Supplement 1): S65–S78.

Klocker, C., and D.L. Strayer. 2004. Interactions among an invasive crayfish (*Orconectes rusticus*), a native crayfish (*Orconectes limosus*) and native bivalves (Sphaeriidae and Unionidae). Northeastern Naturalist 11: 167–178.

Knudsen, K.A., and M.C. Hove. 1997. Spectaclecase (*Cumberlandia monodonta*) conglutinates unique, host(s) elusive. Triannual Unionid Report 11:2.

Kolasa, J. 2002. Microturbellaria. Pages 1–14 *In*: S.D. Rundle, A.L. Robertson, and J.M. Schmid-Araya (eds.). Freshwater meiofauna: biology and ecology. Backhuys, Leiden.

Kolpin, D.W., E.T. Furlong, M.T. Meyer, E.M. Thurman, S.D. Zaugg, L.B. Barber, and H.T. Buxton. 2002. Pharmaceuticals, hormones, and other organic wastewater contaminants in US streams, 1999–2000: A national reconnaissance. Environmental Science and Technology 36: 1202–1211.

Kraemer, L.R. 1970. The mantle flap in three species of *Lampsilis* (Pelecypoda: Unionidae). Malacologia 10: 225–282.

Kryger, J., and H.U. Riisgard. 1988. Filtration rate capacities in 6 species of European freshwater bivalves. Oecologia 77: 34–38.

Kunz, G.F. 1898. The fresh-water pearls and pearl fisheries of the United States. Bulletin of the United States Fish Commission 17:375–426.

Lamouroux, N., and H. Capra. 2002. Simple predictions of instream habitat model outputs for target fish populations. Freshwater Biology 47: 1543–1556.

Lamouroux, N., and I.G. Jowett. 2005. Generalized instream habitat models. Canadian Journal of Fisheries and Aquatic Sciences 62: 7–14.

Lamouroux, N., B. Statzner, U. Fuchs, F. Kohmann, and U. Schmedtje. 1992. An unconventional approach to modelling spatial and temporal variability of local shear stress in stream segments. Water Resources Research 28: 3251–3258.

Lande, R., S. Engen, and B.E. Saether. 1998. Extinction times in finite metapopulation models with stochastic local dynamics. Oikos 83: 383–389.

Layzer, J.B., B. Adair, S. Saha, and L.M. Woods. 2002. Glochidial hosts and other aspects of the life history of the Cumberland pigtoe (*Pleurobema gibberum*). Southeastern Naturalist 2: 73–84.

Layzer, J.B., M.E. Gordon, and R.M. Anderson. 1993. Mussels: The forgotten fauna of regulated rivers. A case study of the Caney Fork River. Regulated Rivers: Research and Management 8: 63–71.

Layzer, J.B., and L.M. Madison. 1995. Microhabitat use by freshwater mussels and recommendations for determining their instream flow needs. Regulated Rivers: Research and Management 10: 329–345.

Lee, C., and M. Hove. 1998. Spectaclecase (*Cumberlandia monodonta*) host(s) still elusive. Triannual Unionid Report 12: 9.

Lefevre, G., and W.C. Curtis. 1911. Metamorphosis without parasitism in the Unionidae. Science 33: 863–865.

Lellis, W.A. 2001. Freshwater mussel survey of the Upper Delaware Scenic and Recreational River: Qualitative survey 2000. Report to the National Park Service. 6 pp. + 5 appendices.

Lellis, W.A., and T.L. King. 1998. Release of metamorphosed juveniles by the green floater, *Lasmigona subviridis*. Triannual Unionid Report 16: 23.

Leopold, L.B., M.G. Wolman, and J.P. Miller. 1964. Fluvial processes in geomorphology. W.H. Freeman, San Francisco. 522 pp.

Levin, S.A. 1992. The problem of pattern and scale in ecology. Ecology 73: 1943–1967.

Lewandowski, K. 1991. Long-term changes in the fauna of family Unionidae bivalves in the Mikołajskie Lake. Ekologia Polska 39: 265–272.

Lewandowski, K. 2006. Bivalves of the family Unionidae in ox-bow lakes of the Bug River. Fragmenta Faunistica 49: 75–79.

Lewandowski, K., and A. Stanczykowska. 1975. The occurrence and role of bivalves of the family Unionidae in Mikołajskie Lake. Ekologia Polska 23: 317–334.

Likens, G.E. (ed.). 1972. Nutrients and eutrophication: The limiting nutrient controversy. Special Publication 1 of the American Society of Limnology and Oceanography. Lawrence, KS. 328 pp.

Linhart, H. and W. Zucchini. 1986. Model selection. John Wiley and Sons, New York. 301 pp.

Lodge, D.M., C.A. Taylor, D.M. Holdich, and J. Skurdal. 2000. Nonindigenous crayfishes threaten North American freshwater biodiversity: lessons from Europe. Fisheries 25(8): 7–20.

Lovett, G.M., C.G. Jones, M.G. Turner, and K.C. Weathers (eds.). 2005. Ecosystem function in heterogeneous landscapes. Springer Verlag, New York. 505 pp.

Lydeard, C., M. Mulvey, and G.M. Davis. 1996. Molecular systematics and evolution of reproductive traits of North American freshwater unionacean mussels (Mollusca: Bivalva) as inferred from 16S rRNA gene sequences. Philosophical Transactions of the Royal Society of London, Series B 351: 1593–1603.

MacIsaac, H.J. 1994. Size-selective predation on zebra mussels (*Dreissena polymorpha*) by crayfish (*Orconectes propinquus*). Journal of the North American Benthological Society 13: 206–216.

MacIsaac, H.J., O.E. Johannsson, J. Ye, W.G. Sprules, J.H. Leach, J.A. McCorquodale, and I.A. Grigorovich. 1999. Filtering impacts of an introduced bivalve (*Dreissena polymorpha*) in a shallow lake: Application of a hydrodynamic model. Ecosystems 2: 338–350.

Madon, S.P., D.W. Schneider, J.A. Stoeckel, and R.E. Sparks. 1998. Effects of inorganic sediment and food concentrations on energetic processes of the zebra mussel, *Dreissena polymorpha*: Implications for growth in turbid rivers. Canadian Journal of Fisheries and Aquatic Sciences 55: 401–413.

Maezono, Y., and T. Miyashita. 2004. Impact of exotic fish removal on native communities in farm ponds. Ecological Research 19: 263–267.

Magnin, E., and A. Stanczykowska. 1971. Quelques donneés sur la croissance, la biomass, et la production annuelle de trois mollusques Unionidae de la region de Montréal. Canadian Journal of Zoology 49: 491–497.

Malmqvist, B., and S. Rundle. 2002. Threats to the running water ecosystems of the world. Environmental Conservation 29:134–153.

Malmqvist, B., R.S. Wotton, and Y.X. Zhang. 2001. Suspension feeders transform massive amounts of seston in large northern rivers. Oikos 92: 35–43.

Marshall, W.B. 1917. *Lampsilis ventricosa cohongoronta* in the Potomac River. The Nautilus 31: 40.

Marshall, W.B. 1918. *Lampsilis ventricosa cohongoronta* in the Potomac valley. The Nautilus 32: 51.

Marshall, W.B. 1930. *Lampsilis ventricosa cohongoronta* in the Potomac River. The Nautilus 44: 19.

Martel, A.L., and J.-S. Lauzon-Guay. 2005. Distribution and density of glochidia of the freshwater mussel *Anodonta kennerlyi* on fish hosts in lakes of the temperate rain forest of Vancouver Island. Canadian Journal of Zoology 83: 419–431.

Martel, A.L., D.A. Pathy, J.B. Magill, C.B. Renaud, S.L. Dean, and S.J. Kerr. 2001. Decline and regional extirpation of freshwater mussels (Unionidae) in a small river system invaded by *Dreissena polymorpha*: The Rideau River, 1993–2000. Canadian Journal of Zoology 79: 2181–2191.

Martell, A., and R. Trdan. 1994. *Venustaconcha ellipsiformis* (Bivalvia: Unionidae), an intermediate host for *Phyllodistomum* (Trematoda: Gorgoderidae) in a Michigan stream. Abstracts from the Annual Meeting of the American Malacological Union, Houston, TX. (original not seen; cited by Dillon 2000).

Martin, G.W., and L.D. Corkum. 1994. Predation of zebra mussels by crayfish. Canadian Journal of Zoology 72:1867–1871.

Master, L. L., B. A. Stein, L. S. Kutner, and G. A. Hammerson. 2000. Vanishing assets: Conservation status of U.S. species. Pages 93–118 in B. A. Stein, L. S. Kutner, and J. S. Adams (eds.). Precious heritage: The status of biodiversity in the United States. Oxford University Press, New York.

McLain, D.C., and M.R. Ross. 2005. Reproduction based on local patch size of *Alasmidonta heterodon* and dispersal by its darter host in the Mill River, Massachusetts, USA. Journal of the North American Benthological Society 24: 139–147.

McLaughlin, R.L., L. Porto, D.L.G. Noakes, J.R. Baylis, L.M. Carl, H.R. Dodd, J.D. Goldstein, D.B. Hayes, and R.G. Randall. 2006. Effects of low-head barriers on stream fishes: Taxonomic affiliations and morphological correlates of sensitive species. Canadian Journal of Fisheries and Aquatic Sciences 63: 766–779.

McMahon, R.F., and A.E. Bogan. 2001. Mollusca: Bivalvia. Pages 331–429 In: J.H. Thorp and A.P. Covich (eds.). Ecology and classification of North American freshwater invertebrates. Second edition. Academic Press, San Diego.

McRae, S.E., J.D. Allan, and J.B. Burch. 2004. Reach- and catchment-scale determinants of the distribution of freshwater mussels (Bivalvia: Unionidae) in south-eastern Michigan, USA. Freshwater Biology 49: 127–142.

Metcalfe-Smith, J.L., J. Di Maio, S.K. Staton, and S.R. DeSolla. 2003. Status of the freshwater mussel communities of the Sydenham River, Ontario, Canada. American Midland Naturalist 150: 37–50.

Metcalfe-Smith, J.L., S.K. Staton, G.L. Mackie, and N.M. Lane. 1998. Changes in the biodiversity of freshwater mussels in the Canadian waters of the lower Great Lakes drainage basin over the past 140 years. Journal of Great Lakes Research 24: 845–858.

Miller, A.C., and B.S. Payne. 1993. Qualitative versus quantitative sampling to evaluate population and community characteristics as a large-river mussel bed. American Midland Naturalist 130: 133–145.

Miller, A.C., and B.S. Payne. 1998. Effects of disturbances on large-river mussel assemblages. Regulated Rivers: Research and Management 14: 179–190.

Miller, A.C., B.S. Payne, and D.W. Aldridge. 1986. Characterization of a bivalve community in the Tangipahoa River, Mississippi. The Nautilus 100: 18–23.

Miller, A.C., B.S. Payne, and R. Tippit. 1992. Characterization of a freshwater mussel (Unionidae) community immediately downriver of Kentucky Lock and Dam in the Tennessee River. Transactions of the Kentucky Academy of Sciences 53: 154–161.

Mitchell, R. 1965. Population regulation of a water mite parasitic on unionid mussels. Journal of Parasitology 51: 990–996.

Mock, K.E., J.C. Brim Box, M.P. Miller, M.E. Downing, and W.R. Hoeh. 2004. Genetic diversity and divergence among freshwater mussel (*Anodonta*) populations in the Bonneville Basin of Utah. Molecular Ecology 13: 1085–1098.

Moore, J.W., J.L. Ruesink, and K.A. McDonald. 2004. Impact of supply-side ecology on consumer-mediated coexistence: Evidence from a meta-analysis. American Naturalist 163: 480–487.

Monroe, E.M., and T.J. Newton. 2001. Seasonal variation in physiological condition of *Amblema plicata* in the Upper Mississippi River. Journal of Shellfish Research 20: 1167–1171.

Morales, Y., L.J. Weber, A.E. Mynett, and T.J. Newton. 2006. Effects of substrate and hydrodynamic conditions on the formation of mussel beds in a large river. Journal of the North American Benthological Society 25: 664–676.

Morris, T.J., and L.D. Corkum. 1996. Assemblage structure of freshwater mussels (Bivalvia: Unionidae) in rivers with grassy and forested riparian zones. Journal of the North American Benthological Society 15: 576–586.

Morrison, J.P.E. 1942. Preliminary report on mollusks found in the shell mounds of the Pickwick Landing basin in the Tennessee River valley. Bulletin of the Bureau of American Ethnology 129:337–392.

Moss, B., P. Johnes, and G. Phillips. 1996. The monitoring of ecological quality and the classification of standing waters in temperate regions: A review and proposal based on a worked scheme for British waters. Biological Reviews 71: 301–339.

Mulcrone, R.S. 2004. Incorporating habitat characteristics and fish hosts to predict freshwater mussel (Bivalvia: Unionidae) distributions in the Lake Erie drainage, southeastern Michigan. Ph.D. dissertation, University of Michigan, Ann Arbor, MI. 139 pp.

Mummert, A.K., R.J. Neves, T.J. Newcomb, and D.S. Cherry. 2003. Sensitivity of juvenile freshwater mussels (*Lampsilis fasciola*, *Villosa iris*) to total and unionized ammonia. Environmental Toxicology and Chemistry 22: 2545–2553.

Myers-Kinzie, M. 1998. The effect of temperature on the respiration of the freshwater mussel *Lampsilis siliquoidea* (Bivalvia: Unionidae). Proceedings of the Indiana Academy of Sciences 107: 85–89.

Myers-Kinzie, M., A. Spacie, C. Rich, and M. Doyle. 2002. Relationship of unionid mussel occurrence to channel stability in urban streams. Verhandlungen der internationale Vereinigung für theoretische und angewandte Limnologie 28: 822–826.

Nagel, K.-O. 2000. Testing hypotheses on the dispersal and evolutionary history of freshwater mussels (Mollusca: Bivalvia: Unionidae). Journal of Evolutionary Biology 13: 854–865.

Naiman, R.J., J.J. Magnuson, D.M. McKnight, and J.A. Stanford (eds.). 1995. The freshwater imperative: A research agenda. Island Press, Washington, DC. 181 pp.

Naimo, T.J. 1995. A review of the effects of heavy metals on fresh-water mussels. Ecotoxicology 4: 341–362.

Nalepa, T.F., and J.M. Gauvin. 1988. Distribution, abundance, and biomass of freshwater mussels (Bivalvia: Unionidae) in Lake St. Clair. Journal of Great Lakes Research 14:411–419.

Nalepa, T.F., B.A. Manny, J.C. Roth, S.C. Mozley, and D.W. Schloesser. 1991. Long-term decline in freshwater mussels (Bivalvia: Unionidae) of the western basin of Lake Erie. Journal of Great Lakes Research 17:214–219.

National Native Mussel Conservation Committee. 1998. National strategy for the conservation of native freshwater mussels. Journal of Shellfish Research 17: 1419–1428.

Nature Serve. 2005. Nature Serve Explorer. http://www.natureserve.org/explorer/servlet/NatureServe?init=Species#anchor_node. Accessed 3 November 2005.

NAWQA. 2006. USGS National Water Quality Assessment Data Warehouse. http://infotrek.er.usgs.gov/traverse/f?p=NAWQA:HOME:6961025307288670248. Accessed 26 July 2006.

Negus, C. L. 1966. A quantitative study of growth and production of unionid mussels in the River Thames at Reading. Journal of Animal Ecology 35: 513–532.

Nepszy, S.J. 1999. The changing fishery regime in Lake Erie. Pages 223–239 *in*: M. Munawar, T. Edsall, and I.F. Munawar (eds.). State of Lake Erie: Past, present and future. Backhuys, Leiden, The Netherlands.

Neves, R.J. (ed.). 1987. Proceedings of the workshop on die-offs of freshwater mussels in the United States. United States Fish and Wildlife Service. 166 pp.

Neves, R.J. 1993. A state-of-the-unionids address. Pages 1–10 *in*: K.S. Cummings, A.C. Buchanan, and L.M. Koch (eds.). Conservation and management of freshwater mussels. Upper Mississippi River Conservation Committee, Rock Island, IL.

Neves, R.J. 1997. Keynote address: a national strategy for the conservation of native freshwater mussels. Pages 1–10 *in*: K.S. Cummings, A.C. Buchanan, C.A. Mayer, and T.J. Naimo (eds). Conservation and management of freshwater mussels II. Initiatives for the future. Upper Mississippi River Conservation Committee, Rock Island, IL.

Neves, R.J. 1999. Conservation and commerce: Management of freshwater mussel (Bivalvia: Unionidae) resources in the United States. Malacologia 41: 461–474.

Neves, R.J., A.E. Bogan, J.D. Williams, S.A. Ahlstedt, and P.W. Hartfield. 1997. Status of aquatic mollusks in the southeastern United States: A downward spiral of diversity. Pages 43–85 *In*: G.W. Benz and D.E. Collins (eds.). Aquatic fauna in peril: The southeastern perspective. Lenz Design and Communications, Decatur, GA.

Neves, R.J., and M.C. Odom. 1989. Muskrat predation on endangered freshwater mussels in Virginia. Journal of Wildlife Management 53: 934–941.

Neves, R.J., and J.C. Widlak. 1988. Occurrence of glochidia in stream drift and on fishes of the upper North Fork Holston River. American Midland Naturalist 119: 111–120.

Newton, T.J. 2003. The effects of ammonia on freshwater unionid mussels. Environmental Toxicology and Chemistry 22: 2543–2544.

Newton, T.J., E.M. Monroe, R. Kenyon, S. Gutreuter, K.I. Welke, and P.A. Thiel. 2001. Evaluation of relocation of unionid mussels into artificial ponds. Journal of the North American Benthological Society 20: 468–485.

Nichols, S.J., and D. Garling. 2000. Food-web dynamics and trophic-level interactions in a multispecies community of freshwater unionids. Canadian Journal of Zoology 78: 871–882.

Nichols, S.J., and D. Garling. 2002. Evaluation of substitute diets for live algae in the captive maintenance of adult and subadult Unionidae. Journal of Shellfish Research 21: 875–881.

Nichols, S.J., H. Silverman, T.H. Dietz, J.W. Lynn, and D.L. Garling. 2005. Pathways of food uptake in native (Unionidae) and introduced (Corbiculidae and Dreissenidae) freshwater bivalves. Journal of Great Lakes Research 31: 87–96.

Nico, L.G., J.D. Williams, and H.L. Jelks. 2005. Black carp: Biological synopsis and risk assessment of an introduced fish. American Fisheries Society Special Publication 32. 337 pp.

Nilsson, C., C. A. Reidy, M. Dynesius, and C. Revenga. 2005. Fragmentation and flow regulation of the world's large river systems. Science 308:405–408.

O'Brien, C.A., and J. Brim Box. 1999. Reproductive biology and juvenile recruitment of the shinyrayed pocketbook, *Lampsilis subangulata* (Bivalvia: Unionidae) in the Gulf Coastal Plain. American Midland Naturalist 142: 129–140.

O'Dee, S.H., and G.T. Watters. 2000. New or confirmed host identifications for ten freshwater mussels. Pages 77–82 *In*: R.A. Tankersley, D.I. Warmolts, G.T. Watters, B.J. Armitage, P.D. Johnson, and R.S. Butler (eds.). Freshwater mollusk symposium proceedings. Ohio Biological Survey, Columbus, OH.

Ohio Division of Water. 2006. Ohio lowhead dam locations map and data. http://www.dnr.state.oh.us/water/dsafety/lowhead_dams/lowhead_locations_map.htm. Accessed 24 Jan 2006.

Ohio EPA. 2006. Ohio EPA Division of Surface Water Biological and Water Quality Reports. http://www.epa.state.oh.us/dsw/document_index/psdindx.html. Accessed 28 July 2006.

Økland, J. 1963. Notes on population density, age distribution, growth, and habitat of *Anodonta piscinalis* Nils. (Moll., Lamellibr.) in a eutrophic Norwegian lake. Nytt Magasin for Zoologi 11:19–43.

Oreskes, N. 2003. The role of quantitative models in science. Pages 13–31 *In*: C.D. Canham, J.J. Cole, and W.K. Lauenroth (eds.). Models in ecosystem science. Princeton University Press, Princeton, NJ.

Ortmann, A.E. 1909. The destruction of the fresh-water fauna in western Pennsylvania. Proceedings of the American Philosophical Society 48:90–110.

Ortmann, A.E. 1912. *Lampsilis ventricosa* (Barnes) in the Upper Potomac drainage. The Nautilus 26: 51–54.

Ortmann, A.E. 1913. The Alleghenian Divide, and its influence upon the freshwater fauna. Proceedings of the American Philosophical Society, Philadelphia, PA. 52:287–390 + 3 plates.

Ortmann, A.E. 1919. A monograph of the naiades of Pennsylvania. Part III: Systematic account of the genera and species. Memoirs of the Carnegie Museum, Pittsburgh, PA. 8: 1–384 + 21 plates.

Ostrovsky, I., M. Gophen, and I. Kalikhman. 1993. Distribution, growth, and ecological significance of the clam *Unio terminalis* in Lake Kinneret, Israel. Hydrobiologia 271: 49–63.

Pace, M.L., J.J. Cole, S.R. Carpenter, and J.F. Kitchell. 1999. Trophic cascades revealed in diverse ecosystems. Trends in Ecology and Evolution 14: 483–488.

Pace, M. L., S. E. G. Findlay, and D. Fischer. 1998. Effects of an invasive bivalve on the zooplankton community of the Hudson River. Freshwater Biology 39:103–116.

Parker, B.C., M.A. Patterson, and R.J. Neves. 1998. Feeding interactions between native freshwater mussels (Bivalvia: Unionidae) and zebra mussels (*Dreissena polymorpha*) in the Ohio River. American Malacological Bulletin 14: 173–179.

Parmalee, P.W., and A.E. Bogan. 1998. The freshwater mussels of Tennessee. University of Tennessee Press, Knoxville, TN. 328 pp.

Paterson, C.G. 1984. A technique for determining apparent selective filtration in the fresh-water bivalve *Elliptio complanata* (Lightfoot). The Veliger 27: 238–241.

Paterson, C.G. 1986. Particle size selectivity in the freshwater bivalve *Elliptio complanata* (Lightfoot). The Veliger 29: 235–237.

Payne, B.S., and A.C. Miller. 2000. Recruitment of *Fusconaia ebena* (Bivalvia: Unionidae) in relation to discharge of the lower Ohio River. American Midland Naturalist 144: 328–341.

Payne, B.S., A.C. Miller, and J. Lei. 1995. Palp to gill area of bivalves: A sensitive indicator of elevated suspended solids. Regulated Rivers: Research and Management 11: 193–200.

Perry, W.L., D.M. Lodge, and G.A. Lamberti. 1997. Impact of crayfish predation on exotic zebra mussels and native invertebrates in a lake-outlet stream. Canadian Journal of Fisheries and Aquatic Sciences 54:120–125.

Perry, W.L., D.M. Lodge, and G.A. Lamberti. 2000. Crayfish (*Orconectes rusticus*) impacts on zebra mussel (*Dreissena polymorpha*) recruitment and algal biomass in a lake-outlet stream. American Midland Naturalist 144: 308–316.

Peters, R.H. 1986. The role of prediction in limnology. Limnology and Oceanography 31: 1143–1159.

Peters, R.H. 1991. A critique for ecology. Cambridge University Press, Cambridge, UK. 366 pp.

Petty, J.T., and G.D. Grossman. 2004. Restricted movement by mottled sculpin (Pisces: Cottidae) in a southern Appalachian stream. Freshwater Biology 49: 631–645.

Polis, G.A., M.E. Power, and G.R. Huxel (eds.). 2004. Food webs at the landscape level. University of Chicago Press. 528 pp.

Pomeroy, L.R., and W.J. Wiebe. 2001. Temperature and substrates as interactive limiting factors for marine heterotrophic bacteria. Aquatic Microbial Ecology 23: 187–204.

Ponyi, J.E. 1992. The distribution and biomass of Unionidae (Mollusca, Bivalvia), and the production of *Unio tumidus* Retzius in Lake Balaton (Hungary). Archiv für Hydrobiologie 125: 245–251.

Poole, K.E., and J.A. Downing. 2004. Relationship of declining mussel biodiversity to stream-reach and watershed characteristics in an agricultural landscape. Journal of the North American Benthological Society 23: 114–125.

Power, M.E. 2006. Environmental controls on food web regimes: a fluvial perspective. Progress in Oceanography 68: 125–133.

Power, M.E., and W.E. Rainey. 2000. Food webs and resource sheds: Towards spatially delimiting trophic interactions. Pages 291–314 In: M.J. Hutchings, E.A. John, and A.J.A. Stewart (eds.). Ecological consequences of habitat heterogeneity. Blackwell, Oxford.

Rabeni, C.F. 1992. Trophic linkage between stream centrarchids and their crayfish prey. Canadian Journal of Fisheries and Aquatic Sciences 49: 1714–1721.

Rahel, F.J. 2000. Homogenization of fish faunas across the United States. Science 288: 854–856.

Raikow, D.F., and S.K. Hamilton. 2001. Bivalve diets in a midwestern US stream: A stable isotope enrichment study. Limnology and Oceanography 46: 514–522.

Randall, R.G., J.R.M. Kelso, and C.K. Minns. 1995. Fish production in fresh waters: Are rivers more productive than lakes? Canadian Journal of Fisheries and Aquatic Sciences 52: 631–643.

Rashleigh, B., and D.L. DeAngelis. 2007. Conditions for coexistence of freshwater mussel species via partitioning of fish host resources. Ecological Modelling 201: 171–178.

Reuling, F.H. 1919. Acquired immunity to an animal parasite. Journal of Infectious Diseases 24: 337–346.

Reynolds, J.D., and R. Donahoe. 2001. Crayfish predation experiments on the introduced zebra mussel, *Dreissena polymorpha*, in Ireland, and their potential for biocontrol. Bulletin Francais de la Peche et de la Pisciculture 361: 669–681.

Ricciardi, A., R.J. Neves, and J.B. Rasmussen. 1998. Impending extinctions of North American freshwater mussels (Unionoida) following the zebra mussel (*Dreissena polymorpha*) invasion. Journal of Animal Ecology 67: 613–619.

Ricciardi, A., F.G. Whoriskey, and J.B. Rasmussen. 1995. Predicting the intensity and impact of *Dreissena* infestation on native unionid bivalves from *Dreissena* field density. Canadian Journal of Fisheries and Aquatic Sciences 52: 1449–1461.

Richardson, J.B., and W.P. Smith. 1994. Nondestructive quantitative sampling for freshwater mussels in variable substrate streams. Proceedings of the Annual Conference of the Southeastern Association of Fish and Wildlife Agencies 48: 357–367.

Rigler, F.H., and R.H. Peters. 1995. Science and limnology. Ecology Institute, Oldendorf, Germany. 239 pp.

Riisgard, H.U. 2001. On measurement of filtration rate in bivalves: The stony road to reliable data. Review and interpretation. Marine Ecology Progress Series 211: 275–291.

Riusech, F.A., and M.C. Barnhart. 2000. Host suitability and utilization in *Venustaconcha ellipsiformis* and *Venustaconcha pleasii* (Bivalvia: Unionidae) from the Ozark Plateaus. Pages 83–91 *In*: R.A. Tankersley, D.I. Warmolts, G.T. Watters, B.J. Armitage, P.D. Johnson, and R.S. Butler (eds.). Freshwater mollusk symposium proceedings. Ohio Biological Survey, Columbus, OH.

Roberts, A.D., and M.C. Barnhart. 1999. Effects of temperature, pH, and CO_2 on transformation of the glochidia of *Anodonta suborbiculata* on fish hosts and in vitro. Journal of the North American Benthological Society 18: 477–487.

Roditi, H.A., N.S. Fisher, and S.A. Sanudo-Wilhelmy. 2000. Uptake of dissolved organic carbon and trace elements by zebra mussels. Nature 407: 78–80.

Rodriguez, M.A. 2002. Restricted movement in stream fish: The paradigm is incomplete, not lost. Ecology 83: 1–13.

Rogers, C.L., and R.V. Dimock. 2003. Acquired resistance of bluegill sunfish *Lepomis macrochirus* to glochidia larvae of the freshwater mussel *Utterbackia imbecillis* (Bivalvia: Unionidae) after multiple infections. Journal of Parasitology 89:51–56.

Rogers, S.O., B.T. Watson, and R.J. Neves. 2001. Life history and population biology of the endangered tan riffleshell (*Epioblasma florentina walkeri*) (Bivalvia: Unionidae). Journal of the North American Benthological Society 20: 582–594.

Rooke, J.B., and G.L. Mackie. 1984. Mollusca of six low-alkalinity lakes in Ontario. Canadian Journal of Fisheries and Aquatic Sciences 41: 777–782.

Rosenberg, D.M., F. Berkes, R.A. Bodaly, R.E. Hecky, C.A. Kelly, and J.M.M. Rudd. 1997. Large-scale impacts of hydroelectric development. Environmental Reviews 5: 27–54.

Saarinen, M., and T. Taskinen. 2004. Aspects of the ecology and natural history of *Paraergasilus rylovi* (Copepoda, Ergasilidae) parasitic in unionids of Finland. Journal of Parasitology 90: 948–952.

Schloesser, D.W., J.L. Metcalfe-Smith, W.P. Kovalak, G.D. Longton, and R.D. Smithee. 2006. Extirpation of freshwater mussels (Bivalvia: Unionidae) following invasion of dreissenid mussels in an interconnecting river of the Laurentian Great Lakes. American Midland Naturalist 155: 307–320.

Schloesser, D.W., T.F. Nalepa, and G.L. Mackie. 1996. Zebra mussel infestations of unionid bivalves (Unionidae) in North America. American Zoologist 36: 300–310.

Schneider, D.W., S.P. Madon, J.A. Stoeckel, and R.E. Sparks. 1998. Seston quality controls zebra mussel (*Dreissena polymorpha*) energetics in turbid rivers. Oecologia 117: 331–341.

Schöne, B.R., E. Dunca, H. Mutvei, and U. Norlund. 2004. A 217-year record of summer air temperature reconstructed from freshwater pearl mussels (*M. margaritifera*, Sweden). Quaternary Science Reviews 23: 1803–1816.

Schwalb, A.N., and M.T. Pusch. 2007. Horizontal and vertical movements of unionid mussels in a lowland river. Journal of the North American Benthological Society 26: 261–272.

Serb, J.M. 2006. Discovery of genetically distinct sympatric lineages in the freshwater mussel *Cyprogenia aberti* (Bivalvia: Unionidae). Journal of Molluscan Studies 72: 425–434.

Shelton, D.N. 1997. Observations on the life history of the Alabama pearl shell, *Margaritifera marrianae* Johnson RI, 1983. Pages 26–29 *in*: Cummings, K.S., A.C. Buchanan, C.A. Mayer, and T.J. Naimo (eds.). Conservation and management of freshwater mussels II: Initiatives for the future. Upper Mississippi River Conservation Commission, Rock Island, IL.

Silverman, H., S.J. Nichols, J.S. Cherry, E. Achberger, J.W. Lynn, and T.H. Dietz. 1997. Clearance of laboratory-cultured bacteria by freshwater bivalves: Differences between lentic and lotic unionids. Canadian Journal of Zoology 75: 1857–1866.

Simpson, C.T. 1896. The classification and geographical distribution of the pearly fresh-water mussels. Proceedings of the United States National Museum, Washington, DC 18: 295–343 + 1 map.

Smith, D.G. 1985. Recent range expansion of the freshwater mussel *Anodonta implicata* and its relationship to clupeid fish restoration in the Connecticut River system. Freshwater Invertebrate Biology 4: 105–108.

Smith, D.R., R.F. Villella, and D. P. Lemarié. 2001. Survey protocol for assessment of endangered freshwater mussels in the Allegheny River, Pennsylvania. Journal of the North American Benthological Society 20: 118–132.

Smith, D.R., R.F. Villella, D. P. Lemarié, and S. von Oettingen. 2000. How much excavation is needed to monitor freshwater mussels? Pages 203–218 *In*: Tankersley, R.A., D.I. Warmolts, G.T. Watters, B.J. Armitage, P.D. Johnson, and R.S. Butler (eds.). Freshwater Mollusk Symposia Proceedings. Ohio Biological Survey, Columbus, OH.

Smith, R.A., R.B. Alexander, and G.E. Schwarz. 2003. Natural background concentrations of nutrients in streams and rivers of the conterminous United States. Environmental Science and Technology 37: 3039–3047.

Smith, V.H. 2003. Eutrophication of freshwater and coastal marine ecosystems. Environmental Science and Pollution Research 10: 1–14.

Soto, D., and G. Mena. 1999. Filter feeding by the freshwater mussel, *Diplodon chilensis*, as a biocontrol of salmon farming eutrophication. Aquaculture 171: 65–81.

Sparks, B.L., and D. L. Strayer. 1998. The effects of low dissolved oxygen on juveniles of *Elliptio complanata* (Bivalvia: Unionidae). Journal of the North American Benthological Society 17: 129–134.

Sparks, D., C. Chafee, and S. Sobiech. 1999. Fish Creek restoration and preservation. Endangered Species Bulletin 24 (1): 12–13.

Steingraeber, M.T., M.R. Bartsch, J.E. Kalas, and T.J. Newton. 2007. Thermal criteria for early life stage development of the winged mapleleaf mussel (*Quadrula fragosa*). American Midland Naturalist 157: 297–311.

Stenroth, R., and P. Nystrom. 2003. Exotic crayfish in a brown water stream: Effects on juvenile trout, invertebrates and algae. Freshwater Biology 48: 466–475.

Stewart, T.W., J.G. Miner, and R.L. Lowe. 1998. An experimental analysis of crayfish (*Orconectes rusticus*) effects on a *Dreissena*-dominated benthic macroinvertebrate community in western Lake Erie. Canadian Journal of Fisheries and Aquatic Sciences 55:1043–1050.

Stone, J., S. Barndt, and M. Gangloff. 2004. Spatial distribution and habitat use of the western pearlshell mussel (*Margaritifera falcata*) in a western Washington stream. Journal of Freshwater Ecology 19: 341–352.

Strayer, D. 1979. Some recent collections of mussels from southeastern Michigan. Malacological Review 12:93–95.

Strayer, D. 1980. The freshwater mussels (Bivalvia: Unionidae) of the Clinton River, Michigan, with comments on man's impact on the fauna, 1870–1978. The Nautilus 94:142–149.

Strayer, D.L. 1981. Notes on the microhabitats of unionid mussels in some Michigan streams. American Midland Naturalist 106:411–415.

Strayer, D. 1983. The effects of surface geology and stream size on freshwater mussel distribution in southeastern Michigan, USA. Freshwater Biology 13:253–264.

Strayer, D.L. 1993. Macrohabitats of freshwater mussels (Bivalvia: Unionacea) in streams of the northern Atlantic Slope. Journal of the North American Benthological Society 12: 236–246.

Strayer, D.L. 1999a. Use of flow refuges by unionid mussels in rivers. Journal of the North American Benthological Society 18: 468–476.

Strayer, D.L. 1999b. Effects of alien species on freshwater mollusks in North America. Journal of the North American Benthological Society 18: 74–98.

Strayer, D.L. 2006. Challenges for freshwater invertebrate conservation. Journal of the North American Benthological Society 25: 271–287.

Strayer, D.L., N.F. Caraco, J.J. Cole, S. Findlay, and M.L. Pace. 1999. Transformation of freshwater ecosystems by bivalves: A case study of zebra mussels in the Hudson River. BioScience 49: 19–27.

Strayer, D.L., Cole, J.J., Likens, G.E., and D.C. Buso. 1981. Biomass and annual production of the freshwater mussel *Elliptio complanata* in an oligotrophic softwater lake. Freshwater Biology 11:435–440.

Strayer, D.L., H. Ewing, and S. Bigelow. 2003. What kinds of spatial and temporal detail are required in models of heterogeneous systems? Oikos 102: 654–662.

Strayer, D.L., and A.R. Fetterman. 1999. Changes in the distribution of freshwater mussels (Unionidae) in the upper Susquehanna River basin, 1955–1997. American Midland Naturalist 142: 328–339.

Strayer, D.L. Hunter, D.C., Smith, L.C., and C. Borg. 1994. Distribution, abundance, and role of freshwater clams (Bivalvia: Unionidae) in the freshwater tidal Hudson River. Freshwater Biology 31: 239–248.

Strayer, D.L., and K.J. Jirka. 1997. The pearly mussels of New York state. Memoirs of the New York State Museum, Albany, NY 26: 1–113 + 27 plates.

Strayer, D.L., and H.M. Malcom. 2007a. Effects of zebra mussels (*Dreissena polymorpha*) on native bivalves: the beginning of the end or the end of the beginning? Journal of the North American Benthological Society 26: 111–122.

Strayer, D.L., and H.M. Malcom. 2007b. Shell decay rates of native and alien freshwater bivalves and implications for habitat engineering. Freshwater Biology 52: 1611–1617.

Strayer, D.L., S.E. May, P. Nielsen, W. Wollheim, and S. Hausam. 1997. Oxygen, organic matter, and sediment granulometry as controls on hyporheic animal communities. Archiv für Hydrobiologie 140: 131–144.

Strayer, D.L., and J. Ralley. 1993. Microhabitat use by an assemblage of stream-dwelling unionaceans (Bivalvia), including two rare species of *Alasmidonta*. Journal of the North American Benthological Society 12: 247–258.

Strayer, D.L., and L.C. Smith. 1996. Relationships between zebra mussels (*Dreissena polymorpha*) and unionid clams during the early stages of the zebra mussel invasion of the Hudson River. Freshwater Biology 36: 771–779.

Strayer, D.L., S. Sprague, and S. Claypool. 1996. A range-wide assessment of populations of *Alasmidonta heterodon*, an endangered freshwater mussel (Bivalvia: Unionidae). Journal of the North American Benthological Society 15: 308–317.

Tankersley, R.A., and R.V. Dimock. 1993. The effect of larval brooding on the filtration rate and particle retention efficiency of *Pyganodon cataracta* (Bivalvia, Unionidae). Canadian Journal of Zoology 71: 1934–1944.

Taskinen, J., T. Makela, and E.T. Valtonen. 1997. Exploitation of *Anodonta piscinalis* (Bivalvia) by trematodes: parasite tactics and host longevity. Annales Zoologici Fennici 34: 37–46.

Taskinen, J., and M. Saarinen. 2006. Burrowing behaviour affects *Paraergasilus rylovi* abundance in *Anodonta piscinalis*. Parasitology 133: 623–629.

Taskinen, J., E.T. Valtonen, and T. Makela. 1994. Quantity of sporocysts and seasonality of 2 *Rhipidocotyle* species (Digenea, Bucephalidae) in *Anodonta piscinalis* (Mollusca, Bivalvia). International Journal for Parasitology 24: 877–886.

Taylor, R.W. 1984. The midwestern naiad *Uniomerus tetralasmus* in West Virginia. The Nautilus 98: 162–164.

Tedla, S., and C.H. Fernando. 1969. Observations on the glochidia of *Lampsilis radiata* (Gmelin) infesting yellow perch, *Perca flavescens* (Mitchill) in the Bay of Quinte, Lake Ontario. Canadian Journal of Zoology 47: 705–712.

Tilman, D., R.M. May, C.L. Lehman, and M.A. Nowak. 1994. Habitat destruction and the extinction debt. Nature 371: 65–66.

Trautman, M.E. 1981. The fishes of Ohio. Second edition. Ohio State University Press, Columbus, OH. 782 pp.

Trdan, R.J. 1981. Reproductive biology of *Lampsilis radiata siliquoidea* (Pelecypoda: Unionidae). American Midland Naturalist 106: 243–248.

Trimble, S.W. 1981. Changes in sediment storage in the Coon Creek basin, Driftless Area, Wisconsin, 1853 to 1975. Science 214: 181–183.

Tucker, M.E. 1937. Studies on the life cycles of two species of fresh-water mussels belonging to the genus *Anodonta*. Biological Bulletin 54: 117–124.

Tudorancea, C. 1972. Studies on Unionidae populations from the Crapina and Jijila complex of pools (Danube zone liable to inundation). Hydrobiologia 39: 527–561.

Turner, M.G., R.H. Gardner, and R.V. O'Neill. 2001. Landscape ecology in theory and practice. Springer-Verlag, New York. 401 pp.

Tyrrell, M., and D.J. Hornbach. 1998. Selective predation by muskrats on freshwater mussels in 2 Minnesota rivers. Journal of the North American Benthological Society 17: 301–310.

University of Michigan Museum of Zoology. 2005. Catalogue searches. www.lsa.umich.edu/ummz/areas/fish/newfishsearch.html. Accessed 8 December 2005.

USFWS. 2002. Certus chemical spill natural resource damage assessment: Initial restoration and compensation determination plan. United States Fish and Wildlife Service, Region 5, Gloucester, VA. (Available at http://unionid.smsu.edu/Documents/Draft.Certus.RCDP.pdf).

USFWS. 2007. Species information: Threatened and endangered animals and plants. Clams. http://ecos.fws.gov/tess_public/SpeciesReport.do?groups=F&listingType=L&mapstatus=1. Accessed 28 May 2007.

USGS. 2006a. USGS water quality data for the nation. http://waterdata.usgs.gov/nwis/qw. Accessed 26 July 2006.

USGS. 2006b. Upper Midwest Environmental Sciences Center Water Quality Data. http://www.umesc.usgs.gov/data_library/water_quality/water_quality_page.html. Accessed 28 July 2006.

Vandermeer, J.H., and D.E. Goldberg. 2003. Population ecology: First principles. Princeton University Press, Princeton, NJ. 280 pp.

Vanderploeg, H.A., J.R. Liebig, and T.F. Nalepa. 1995. From picoplankton to microplankton: Temperature-driven filtration by the unionid bivalve *Lampsilis radiata siliquoidea* in Lake St. Clair. Canadian Journal of Fisheries and Aquatic Sciences 52: 63–74.

van der Schalie, H. 1938. The naiad fauna of the Huron River in southeastern Michigan. Miscellaneous Publications of the University of Michigan Museum of Zoology 40: 1–83.

van der Schalie, H. 1945. The value of mussel distribution in tracing stream confluence. Papers of the Michigan Academy of Science, Arts, and Letters 30: 355–373.

van der Schalie, H. 1970. Hermaphroditism among North American freshwater mussels. Malacologia 10: 93–110.

van der Schalie, H., and A. van der Schalie. 1950. The mussels of the Mississippi River. American Midland Naturalist 44: 448–466.

van der Schalie, H., and A. van der Schalie. 1963. The distribution, ecology, and life history of the mussel, *Actinonaias ellipsiformis* (Conrad), in Michigan. Occasional Papers of the University of Michigan Museum of Zoology, Ann Arbor, MI 633: 1–17.

Vannote, R.L., and G.W. Minshall. 1982. Fluvial processes and local lithology controlling abundance, structure, and composition of mussel beds. Proceedings of the National Academy of Sciences 79: 4103–4107.

van Snik Gray, E., W.A. Lellis, J.C. Cole, and C.S. Johnson. 2002. Host identification for *Strophitus undulatus* (Bivalvia: Unionidae), the creeper, in the upper Susquehanna River basin, Pennsylvania. American Midland Naturalist 147: 153–161.

Vaughn, C.C. 1993. Can biogeographic models be used to predict the persistence of mussel populations in rivers? Pages 117–122 *In*: K.S. Cummings, A.C. Buchanan, and L.M. Koch (eds.). Conservation and management of freshwater mussels. Upper Mississippi River Conservation Committee, Rock Island, IL.

Vaughn, C.C. 1997. Regional patterns of mussel species distributions in North American rivers. Ecography 20: 107–115.

Vaughn, C.C. 2000. Changes in the mussel fauna of the middle Red River drainage: 1910–present. Pages 225–232 *In*: Tankersley, R.A., D.I. Warmolts, G.T. Watters, B.J. Armitage, P.D. Johnson, and R.S. Butler (eds.). Freshwater Mollusk Symposia Proceedings. Ohio Biological Survey, Columbus, OH.

Vaughn, C.C., K.B. Gido, and D.E. Spooner. 2004. Ecosystem processes performed by unionid mussels in stream mesocosms: species roles and effects of abundance. Hydrobiologia 527: 35–47.

Vaughn, C.C., and C.C. Hakenkamp. 2001. The functional role of burrowing bivalves in freshwater ecosystems. Freshwater Biology 46: 1431–1446.

Vaughn, C.C., and M. Pyron. 1995. Population ecology of the endangered Ouachita rock-pocketbook mussel, *Arkansia wheeleri* (Bivalvia: Unionidae), in the Kiamichi River, Oklahoma. American Malacological Bulletin 11: 145–151.

Vaughn, C.C., and D.E. Spooner. 2004. Status of the mussel fauna of the Poteau River and implications for commercial harvest. American Midland Naturalist 152: 336–346.

Vaughn, C.C., and D.E. Spooner. 2006. Unionid mussels influence macroinvertebrate assemblage structure in streams. Journal of the North American Benthological Society 25: 691–700.

Vaughn, C.C., D.E. Spooner, and H.S. Galbraith. 2007. Context-dependent species identity effects within a functional group of filter-feeding bivalves. Ecology 88: 1654-1662.

Vaughn, C.C., and C.M. Taylor. 2000. Macroecology of a host-parasite relationship. Ecography 23: 11–20.

Vaughn, C.C., D.E. Spooner, and B.W. Hoagland. 2002. River weed growing epizootically on freshwater mussels. Southwestern Naturalist 47: 604–605.

Vidrine, M.F., and J.L. Wilson. 1991. Parasitic mites (Acari, Unionicolidae) of fresh-water mussels (Bivalvia, Unionidae) in the Duck and Stones Rivers in central Tennessee. Nautilus 105: 152–158.

Villella, R.F., D.R. Smith, and D.P. Lemarié. 2004. Estimating survival and recruitment in a freshwater mussel population using mark-recapture techniques. American Midland Naturalist 151: 114–133.

Vitousek, P.M. 1994. Beyond global warming: Ecology and global change. Ecology 75: 1861–1876.

Vitousek, P.M., J.D. Aber, R.W. Howarth, G.E. Likens, P.A. Matson, D.W. Schindler, W.H. Schlesinger, and D. Tilman. 1997. Human alteration of the lobal nitrogen cycle: Sources and consequences. Ecological Applications 7: 737–750.

Wächtler, K., M.C. Dreher-Mansur, and T. Richter. 2001. Larval types and early postlarval biology in naiads (Unionoida). Pages 93–125 *in*: G. Bauer and K. Wächtler (eds.). Ecology and evolution of the freshwater mussels Unionoida. Springer-Verlag, Berlin.

Walker, J.M., J.P. Curole, D.E. Wade, E.G. Chapman, A.E. Bogan, G.T. Watters, and W.R. Hoeh. 2006. Taxonomic distribution and phylogenetic utility of gender-associated mitochondrial genomes in the Unionoida (Bivalvia). Malacologia 48: 265–282.

Walker, K.F., M. Byrne, C.W. Hickey, and D.S. Roper. 2001. Freshwater mussels (Hyriidae) of Australasia. Pages 5–31 *in*: G. Bauer and K. Wächtler (eds.). Ecology and evolution of the freshwater mussels Unionoida. Springer–Verlag, Berlin.

Wallace, J.B., and R.W. Merritt. 1980. Filter-feeding ecology of aquatic insects. Annual Review of Entomology 25: 103–132.

Warren, M.L., and W.R. Haag. 2005. Spatio-temporal patterns of the decline of freshwater mussels in the Little South Fork Cumberland River, USA. Biodiversity and Conservation 14: 1383–1400.

Waters, T.F. 1995. Sediment in streams: Sources, biological effects and control. American Fisheries Society Monograph 7: 1–251.

Watters, G.T. 1990. 1990 survey of the unionids of the Big Darby Creek system. Final Report to The Nature Conservancy, Ohio Chapter. 36 pages + Appendices A–C.

Watters, G.T. 1992. Unionids, fish, and the species-area curve. Journal of Biogeography 19: 481–490.

Watters, G.T. 1993. A guide to the freshwater mussels of Ohio. Third edition. Ohio Department of Natural Resources, Columbus OH. 106 pp.

Watters, G.T. 1994a. Clubshell (*Pleurobema clava*) and northern riffleshell (*Epioblasma torulosa rangiana*) recovery plan. United States Fish and Wildlife Service, Hadley, MA. 68 pp.

Watters, G.T. 1994b. Sampling freshwater mussel populations: The bias of muskrat middens. Walkerana 7: 63–69.

Watters, G.T. 1996. Small dams as barriers to freshwater mussels (Bivalvia, Unionidae) and their hosts. Biological Conservation 75: 79–85.

Watters, G.T. 1997. Glochidial metamorphosis of the freshwater mussel *Lampsilis cardium* (Bivalvia: Unionidae) on larval tiger salamanders, *Ambystoma tigrinum* ssp. (Amphibia: Ambystomidae). Canadian Journal of Zoology 75: 505–508.

Watters, G.T. 1999. Morphology of the conglutinate of the kidneyshell freshwater mussel *Ptychobranchus fasciolaris*. Invertebrate Biology 118: 289–295.

Watters, G.T. 2000. Freshwater mollusks and water quality: A review of the effects of hydrologic and instream habitat alterations. Pages 261–274 *In*: Tankersley, R.A., D.I. Warmolts, G.T. Watters, B.J. Armitage, P.D. Johnson, and R.S. Butler (eds.). Freshwater Mollusk Symposia Proceedings. Ohio Biological Survey, Columbus, OH.

Watters, G.T. 2002. The kinetic conglutinate of the creeper freshwater mussel *Strophitus undulatus* (Say, 1817). Journal of Molluscan Studies 68: 155–158.

Watters, G.T., T. Menker, S. Thomas, and K. Kuehnl. 2005. Host identifications or confirmations. Ellipsaria 7(2): 11–12.

Watters, G.T., and S.H. O'Dee. 1997. Identification of potential hosts. Triannual Unionid Report 13: 38–39.

Watters, G.T., and S.H. O'Dee. 1998. Metamorphosis of freshwater mussel glochidia (Bivalvia: Unionidae) on amphibians and exotic fishes. American Midland Naturalist 139: 49–57.

Watters, G.T., and S.H. O'Dee. 1999. Glochidia of the freshwater mussel Lampsilis overwintering on fish hosts. Journal of Molluscan Studies 65: 453–459.

Weaver, L.R., G.B. Pardue, and R.J. Neves. 1991. Reproductive biology and fish hosts of the Tennessee clubshell *Pleurobema oviforme* (Mollusca: Unionidae) in Virginia. American Midland Naturalist 126: 82–89.

Weinstein, J.E., and K.D. Polk. 2001. Phototoxicity of anthracene and pyrene to glochidia of the freshwater mussel *Utterbackia imbecillis*. Environmental Toxicology and Chemistry 20: 2021–2028.

Weir, G.P. 1977. An ecology of the Unionidae in Otsego Lake with special references to immature stages. Biological Field Station of the State University College at Oneonta, Oneonta, NY, Occasional Paper 4. 108 pp.

Weiss, J.L., and J.B. Layzer. 1995. Infestations of glochidia on fishes in the Barren River Kentucky. American Malacological Bulletin 11: 153–159.

Welker, M., and N. Walz. 1998. Can mussels control the plankton in rivers? A planktological approach applying a Lagrangian sampling strategy. Limnology and Oceanography 43: 753–762.

Wetzel, R.G. 2001. Limnology: Lake and river ecosystems. Third edition. Academic Press, San Diego, CA. 1006 pp.

Wildish, D., and D. Kristmanson. 1997. Benthic suspension feeders and flow. Cambridge University Press, Cambridge, UK. 409 pp.

Williams, J. D., M. L. Warren, K. S. Cummings, J. L. Harris, and R. J. Neves. 1993. Conservation status of the freshwater mussels of the United States and Canada. Fisheries 18(9): 6–22.

Wilson, A.J., J.A. Hutchings, and M.M. Ferguson. 2004a. Dispersal in a stream-dwelling salmonid: Inferences from tagging and microsatellite studies. Conservation Genetics 5: 25–37.

Wilson, K.A., J.J. Magnuson, D.M. Lodge, A.M. Hill, T.K. Kratz, W.L. Perry, and T.V. Willis. 2004b. A long-term rusty crayfish (*Orconectes rusticus*) invasion: Dispersal patterns and community change in a north temperate lake. Canadian Journal of Fisheries and Aquatic Sciences 61: 2255–2266.

Yeager, B.L., and C.F. Saylor. 1995. Fish hosts for four species of freshwater mussels (Pelecypoda: Unionidae) in the Upper Tennessee River drainage. American Midland Naturalist 133: 1–6.

Yeager, M.M., D.S. Cherry, and R.J. Neves. 1994. Feeding and burrowing behaviors of juvenile rainbow mussels, *Villosa iris* (Bivalvia, Unionidae). Journal of the North American Benthological Society 13: 217–222.

Young, M.R., P.J. Cosgrove, and L.C. Hastie. 2001. The extent of, and causes for, the decline of a highly threatened naiad: *Margaritifera margaritifera*. Pages 337–357 *in*: G. Bauer and K. Wächtler (eds.). Ecology and evolution of the freshwater mussels Unionoida. Springer-Verlag, Berlin.

Young, M.R., and J.C. Williams. 1983. Redistribution and local recolonisation by the freshwater pearl mussel *Margaritifera margaritifera* (L.). Journal of Conchology 31: 225–234.

Young, M., and J. Williams. 1984a. The reproductive biology of the freshwater pearl mussel *Margaritifera margaritifera* (Linn.) in Scotland I. Field studies. Archiv für Hydrobiologie 99: 405–422.

Young, M., and J. Williams. 1984b. The reproductive biology of the freshwater pearl mussel *Margaritifera margaritifera* (Linn.) in Scotland II. Laboratory studies. Archiv für Hydrobiologie 100: 29–43.

Zahner-Meike, E., and J.M. Hanson. 2001. Effect of muskrat predation on naiads. Pages 163–184 *In*: G. Bauer and K. Wächtler (eds.). Ecology and evolution of the freshwater mussels Unionoida. Springer-Verlag, Berlin.

Zale, A.V., and R.J. Neves. 1982a. Fish hosts of four species of lampsiline mussels (Mollusca: Unionidae) in Big Moccasin Creek, Virginia. Canadian Journal of Zoology 60: 2535–2542.

Zale, A.V., and R.J. Neves. 1982b. Identification of a fish host for *Alasmidonta minor* (Mollusca: Unionidae). American Midland Naturalist 107: 386–388.

Zimmerman, G.F., and F.A. de Szalay. 2007. Influence of unionid mussels (Mollusca: Unionidae) on sediment stability: An artificial stream study. Archiv für Hydrobiologie 168: 299–306.

Zimmerman, L.L., and R.J. Neves. 2002. Effects of temperature on duration of viability for glochidia of freshwater mussels (Bivalvia: Unionidae). American Malacological Bulletin 17: 31–35.

Zimmerman, L.L., R.J. Neves, and D.G. Smith. 2003. Control of predacious flatworms *Macrostomum* sp. in culturing juvenile freshwater mussels. North American Journal of Aquaculture 65: 28–32.

Ziuganov, V., A. Zotin, L. Nezlin, and V. Tretiakov. 1994. The freshwater pearl mussels and their relationships with salmonid fish. VNIRO Publishing House, Moscow. 104 pages.

INDEX

Acid mine drainage, 56
Alasmidonta heterodon, glochidial prevalence on hosts, 74
Alasmidonta viridis
　glochidial prevalence on hosts, 74
　range boundaries, 26–27
Alien species, effects on unionoids, 20, 94–95, 108, 110, 115–116
Allee effects, 41
Ammonia, 56–59, 62, 64, 117, 153
Amphibians, as hosts, 66
Anodonta
　conservation status, 19
　range expansion, 20
Anodonta anatina, filtration rate, 89
Anodonta cygnea, glochidial prevalence on hosts, 74
Anodonta implicata
　limited by host population, 84
　competition with zebra mussels, 95
Anodonta kennerlyi, host use, 70–71
Anodonta piscinalis, glochidial prevalence on hosts, 74

Bacteria, as food, 88, 91
Big Darby Creek, Ohio, 32, 114–115
Big Moccasin Creek, Virginia, 75
Biodeposition, effects on juveniles, 64, 130
Black carp (*Mylopharyngodon piceus*), 114–116

Calcium, 53–54
Cellulase, 90
Chlorine, 56
Classification of unionoids, 10–13
Climate
　change and unionoids, 30, 55, 61
　effects on geographic ranges, 26–27, 61
Clinton River, Michigan, 18, 32, 150
Competition
　apparent, 131
　for food, 98–103
　interspecific, 25, 73, 90, 92, 94–95
　intraspecific, 72–73
　with insects, 25, 103
Connecticut River, Massachusetts, 84
Conservation, 15–21
　range thinning, 16–18
Conservation status, 15–16, 19
　body size effects, 17, 19
　host fish effects, 68–69

Conservation status, *continued*
    phylogenetic patterns, 16, 17, 19
Copepods, as parasites, 108–109
Corbicula, 116, 131
Crayfish, as predators, 55, 108, 110, 116
Cumberlandia monodonta
    hosts of, 66–67
    photograph of dense bed, 63
Cumberland River, 128
Current speed, 44, 48
    role in food delivery, 52–53, 103
Cyclonaias tuberculata, habitat, 44

Dams
    effects on food resources, 97
    effects on metapopulations, 32, 37
    effects on unionoids, 20, 32, 50, 55, 84, 113–115
Demographic analyses, 96–97, 111, 117, 129, 155
Density dependence, of
    fertilization, 41
    glochidial survival, 70, 73
    habitat quality, 62–64
    in models of host use, 80–84
    predation, 107
Deposit feeding, 88, 90, 91
Detritus, as food, 87–88
Diseases, 109
Dispersal, 25–41, 117
    barriers to, 26–30
    interaction with other controlling factors, 41, 129–130
    rates, 31, 35
    research needs, 39–41
    role of host fishes, 31, 35–36
Dissolved organic matter, as food, 88, 90
Dreissena. *See* Zebra mussel
Droughts, 50, 54, 61

Ecological importance of unionoids, 9, 92
Economic importance of unionoids, 10
Eel River, California, 52

Elliptio complanata
    effects of endocrine disruptors, 59
    fertilization success, 41
    competition with zebra mussels, 95
Elliptio dilatata, microhabitat, 46
Empirical models, 122, 133–142, 144, 152–154
Endocrine disruptors, 58–60
Epioblasma
    conservation status, 19
    host use, 68
Erie Canal, 29
Erie, Lake, 114–115
Etherioidea, 10–11
Eutrophication
    effects on unionoids, 54, 57, 78, 80, 115, 118
    extent, 98
    growth rate effects, 96
Extinction, 15, 19
Extinction debt, 31, 34, 35, 39

Fatty acids, in diet, 90
Filtration rates, 89
Fisheries for unionoids, 10, 109–110
Fishes. *See also* Host fishes
    as predators, 55, 108
Flatworms, as predators, 108
Food and feeding, 87–104
    bacteria in diet, 88, 91
    cellulase, 90
    deposit feeding, 88, 90, 91
    detritus in diet, 87–88
    diet of unionoids, 87–92
    dissolved organic matter, 88, 90
    effects on resource levels, 98–103
    fatty acids in diet, 90
    fecundity control by food, 93, 96
    filtration rates, 89
    interspecific differences, 89, 90
    limitation by food, 92–104, 118, 130–131
    ontogenetic shifts, 89

phytoplankton in diet, 87–88
plasticity, 90, 92
research needs, 104
size selection, 88–89
stable isotope studies, 91–92
zooplankton in diet, 87–88, 92
Food limitation
caused by unionoid populations, 98–103
conditions for, 92–104
environmental conditions as modulators, 103
Frankenstein's monster, as metaphor, 3–4, 122
Freshwater drum (*Aplodinotus grunniens*), 131
Freshwater mussels. *See* Unionoida; Zebra mussels
Fusconaia ebena
limited by host population, 84
recruitment patterns, 48
Fusconaia flava, range boundaries, 26–27, 29

Geographic ranges, 25–30
Glochidia, 13
prevalence rates on hosts, 73–76
survival, 70–72
Grand River, Ohio, 32
Growth rates
effects of food, 52–53, 96
effects of temperature, 54–55

Habitat, 43–64
as limiting factor, 43, 61, 129–130
critique of traditional views, 43–47
density-dependent effects on quality, 62–64
functional approach, 48–61
loss, effects on metapopulations, 35–37
models of, 146–149
"mussel's eye" approach, 48
research needs, 60, 61, 64, 117

Hallwil, Lake, Switzerland, 96–97
Hamiota, host use, 68
Hamiota subangulata, host use, 70
Heavy metals. *See* metals
Hermaphroditism, 13, 41
Heterogeneity, spatial, 123–124, 126–128
Heterogeneity, temporal, 123–124, 128–129
Hog Creek, Ohio, 114–115
Holston River, Virginia, 106
Host fishes, 65–86
attraction by mussels, 14
effects on conservation status, 68–69
human effects on, 85, 113–114, 131
intraspecific variation in use, 68, 70
limiting unionoid populations in nature, 84–86, 118, 130–131
models of effects on unionoid populations, 76–84, 149–151
phylogenetic patterns of use, 66–67
research needs, 86
specificity, 66–70, 86
use in nature, 70–71
Hudson River, New York, 94–95
Huron River, Michigan, 47
Hydraulic models, 50, 52
Hyridella depressa, food limitation, 96
Hyriidae, classification, 10–13

Igor the hunchback, correct name, 4
Immune response, of fish hosts, 67–68, 70–73, 86, 117
cross-species immunity, 72
effect of host size and age, 71–73
effect of temperature, 55
models of, 76–84
Insects, as competitors, 25, 103
Interactions (among controlling factors), 41, 117, 121, 123, 129–132
in empirical models, 139–140
in mechanistic models, 132
Invasive species. *See* Alien species

James River, Missouri, 75
Juvenile mussels
    ammonia toxicity, 56–57
    biodeposition effects, 64
    calcium requirements, 54
    food and feeding, 90
    food supply, 52, 64
    habitat, 48
    oxygen requirements, 53
    predation, 55
    sediment-bound pollutants, 58

Kuivasjärvi, Lake, Finland, 74

Lag effects. *See* Time lags
Lampsilis, conservation status, 19
Lampsilis cardium
    introduction to Potomac River basin, 29–30
    lure to attract hosts, 5
Lampsilis fasciola
    range boundaries, 28
    range thinning 18
Lampsilis radiata, glochidial prevalence on hosts, 75
Lampsilis reeveiana brevicula, immunity in hosts, 71–72
Lampsilis siliquoidea
    glochidial prevalence on hosts, 75
    size-selectivity in diet, 89
Land use effects, 50–51
Larvae. *See* Glochidia
Lasmigona complanata, range boundaries, 27
Lasmigona subviridis, life history, 65
Legacies. *See* Time lags
Liebig's Law of the Minimum, 123–126, 143
Life history of unionoids, 13–14, 65–66
Life span of unionoids, 14
Life table analysis. *See* Demographic analyses

Ligumia ochracea, competition with zebra mussels, 95

Margaritifera auricularia, conservation status, 17
Margaritifera hembeli
    growth rates, 52
    sediment requirements, 49
Margaritifera margaritifera
    effect of temperature, 55, 61
    food limitation, 96, 97
    infestation rates on hosts, 73–76
    limitation by hosts, 85
    sediment requirements, 49, 57
Margaritiferidae, classification, 10–12
Mechanistic models, 122, 132–133, 143, 146–152
Medionidus conradicus, glochidial prevalence on hosts, 75
Metals, 56–58
Metapopulations, 30–41
    habitat loss effects, 35–37
    human impacts on, 32–39
    models, 33–41
    research needs, 39–41
    stochastic effects, 39
Mill River, Massachusetts, 74
Mississippi River, 84, 109, 128
Mites, as parasites, 109
Models
    domains of, 125, 144, 146
    empirical, 122, 133–142, 144, 152–154
    examples of integrative, 146–154
    food depletion by unionoids, 98–103
    fish hosts, 76–84
    mechanistic, 122, 132–133, 143, 146–152
    metapopulation, 33–41
    predictive power of, 133, 144–145
    strategies for developing, 144–146, 154–156

Muskrats, as predators, 105–107, 110

Narrow Lake, Alberta, 74
Necturus maculosus (mudpuppy), as host, 66
Neversink River, New York, 45, 51, 149
Nitrate, 59, 153

Obliquaria reflexa, life history, 65
Ohio River, 48
Ontario, Lake, 75
Oxygen, 53–54, 57, 62, 64

Parasites, 108–109, 118
Pearly mussels. *See* Unionoida
Pembina River, Alberta, 57
Pesticides, 56–58
Phytoplankton, as food, 87–88
Placopecten magellanicus (marine giant scallop), 53
Pleurobema, host use, 68
Pleurobema clava, range thinning, 17
Pleurobema oviforme, glochidial prevalence on hosts, 75
Pollution, 20, 54, 56–60, 117
　acid mine drainage, 56
　ammonia, 56–59, 62, 64, 117, 153
　chlorine, 56
　endocrine disruptors, 58–60
　eutrophication, 54, 57, 78, 80, 96, 98, 115, 118
　metals, 56–58
　pesticides, 56–58
Population density, in nature, 9, 100
Potomac River, Maryland, 29
Predators and predation, 105–111, 117, 130–131
　crayfish, 55, 108, 110
　fishes, 55, 108
　flatworms, 108

　humans, 109–110
　invertebrates, 108
　muskrats, 105–107, 110
　optimal foraging models, 106
　refuges from, 55, 107
　research needs, 110–111
Pyganodon
　conservation status, 19
　range expansion, 20
Pyganodon grandis
　glochidial prevalence on hosts, 74
　hosts, 67
　range boundaries, 29

Raisin River, Michigan, 18, 37
Range boundaries, 25–30
Range thinning, 16–18
Reductionism, shortcomings, 3
Research needs, 117–118
　dispersal, 39–41, 117
　food and feeding, 104, 118
　habitat, 60, 61, 64, 117
　host fishes, 86, 117–118
　parasites and predators, 110–111, 117–118

Sediments
　grain size, 44
　limitation by, 49–51, 57
　penetrability, 49, 64
　stability, 50–52, 61, 64, 117
Shannon Lake, Minnesota, 75
Shear stress, 48–49
Shoulder of Mutton Pond, England, 74
Silt and siltation, 57, 103
Simpsonaias ambigua, life history, 66
"Sink" habitats, 127
South River, Nova Scotia, 74
Spree, River, Germany, 92
Spring River, Missouri, 75
Stable isotope studies of diet, 91–92
Stac Burn, Scotland, 74

St. Clair, Lake, 89
Stream size, as habitat factor, 46–47
Strophitus undulatus
  hosts, 67
  life history, 65
Systematics. *See* Classification

Temperature
  as limiting factor, 54–55
  importance in ammonia toxicity, 56–57
Time lags, 20–21, 84, 85, 128
  in metapopulations, 31, 39, 54
Toxins. *See* Pollution
Toxolasma, range expansion, 20
Trematodes, as parasites, 108
Trophic cascade, 121–122
Tucker Pond, Rhode Island, 96
Uniomerus, range expansion, 20

Unionidae, classification, 10–13
Unionoida, classification, 10–13
  number of species, 11
  population density, 9
Unionoidea, classification, 10–13
Unio tumidus, food limitation in, 96

Utterbackia
  conservation status, 19
  life history, 65

Velesunio ambiguus, diet, 91
Venustaconcha ellipsiformis
  glochidial prevalence on hosts, 75
  range boundaries, 28
Villosa nebulosa, glochidial prevalence on hosts, 75
Villosa vanuxemi, glochidial prevalence on hosts, 75

Webatuck Creek, New York, 45, 51, 124
Worden Pond, Rhode Island, 96

Yawgoo Pond, Rhode Island, 96

Zebra mussel (*Dreissena polymorpha*)
  dissolved organic matter in diet, 88, 90
  effects on unionoids, 20, 94–95
  effects on habitat quality, 62
Zoogeographic regions, 26
Zooplankton, as food, 87–88, 92

Compositor:	Michael Bass Associates
Text:	10.75/14 Bembo
Display:	Bembo
Printer and Binder:	Thomson-Shore, Inc.